Systems & Control: Foundations & Applications

More information about this series at http://www.springer.com/series/4895

Ziyang Meng • Tao Yang • Karl H. Johansson

Modelling, Analysis, and Control of Networked Dynamical Systems

 Birkhäuser

Ziyang Meng
Department of Precision Instrument
Tsinghua University
Beijing, China

Tao Yang
State Key Laboratory of Synthetical
Automation for Process Industries
Northeastern University
Shenyang, China

Karl H. Johansson
School of Electrical Engineering
and Computer Science
KTH Royal Institute of Technology
Stockholm, Sweden

ISSN 2324-9749   ISSN 2324-9757   (electronic)
Systems & Control: Foundations & Applications
ISBN 978-3-030-84684-8   ISBN 978-3-030-84682-4   (eBook)
https://doi.org/10.1007/978-3-030-84682-4

Mathematics Subject Classification: 93A16, 93A14, 93A15, 93D15

This book is published under the imprint Birkhäuser, www.birkhauser-science.com by the registered company Springer Nature Switzerland AG
The registered company address is: Gewerbestrasse 11, 6330 Cham, Switzerland

My parents, Wei Meng and Li Guo
My wife, Qian Wang
My daughter, Qinglin Meng

— Ziyang Meng

My parents, Shaokun Yang and Ying Xing

— Tao Yang

My parents, Sigurd and Gunvor
My wife, Liselott
My sons, Kasper and Felix

— Karl H. Johansson

Preface

Rapid developments in digital systems, communication, and sensing technologies have led to the emergence of networked dynamical systems. These systems consist of a large number of interconnected subsystems (agents), which are required to cooperate in order to achieve a desirable global objective. Potential applications for such networked systems can be found in biology, social science, computer science, and engineering. As many applications are components of our infrastructure, how to control and coordinate agents in such networked systems is of great interest.

The revolutionary idea is that by carefully designing distributed control laws, a group of autonomous agents can achieve a collective task by cooperatively and locally exchanging information. Interactions are local in the sense that an agent can only interact with a subset of agents. Under such a distributed framework, this book aims to provide some recent results on modelling, analysis, control, and applications of networked dynamical systems. In particular, from a theoretical perspective, it is valuable to analyze some fundamental properties of emergent behaviors of networked dynamical systems. Typical emergent behaviors are observed in circuit networks, social networks, and biological networks. Precise mathematical models are introduced to reconstruct these group behaviors including synchronizing to a time-varying trajectory, clustering into different groups, flocking with alignment, separation, and cohesion properties. From a methodology perspective, it is essential to model networked dynamical systems in a realistic environment and to properly design their controls such that the overall objective is achieved even when there are only limited resources available. Systems with a variety of constraints are considered in this book. Typical constraints include those on dynamics, energy, communication, sensing, and control. From an application perspective, it is indispensable to implement the proposed distributed control algorithms into practical systems and adjust them depending on various engineering requirements. Typical applications include multiple unmanned airborne vehicles, spacecraft formation flying, heavy-duty vehicle platooning, smart grid, and social networks.

This book consists of four parts. The first part presents preliminaries in Chap. 2 and the general networked dynamical model in Chap. 3 that will be repeatedly used in the rest of the book. The second part focuses on the behavior analysis of networked dynamical systems. In particular, synchronization of network dynamical systems is discussed in Chap. 4 and the extension to synchronization with non-expansive dynamics is studied in Chap. 5. In addition, periodic solutions of networked dynamical systems and modulus consensus of cooperative-antagonistic network are discussed in Chaps. 6 and 7, respectively. The third part studies control problems for networked dynamical systems. We particularly solve control problems with input constraint in Chap. 8, large delays in Chap. 9, and heterogenous dynamics in Chap. 10. The last part presents three applications of networked dynamical systems in Chaps. 11, 12, and 13, where the cooperative attitude control problem of spacecraft formation flying, the rendezvous problem of multi-robot systems, and the energy resource coordination problem of power network are studied, respectively.

We would like to thank our colleagues and collaborators for their contributions to the works that form the basis of this book. In particular, we are indebted to Professors Wei Ren, Zongli Lin, Dimos V. Dimarogonas, Yiguang Hong, Ming Cao, Brian D. O. Anderson, and Sandra Hirche for many useful discussions and professional advice. We thank Mr. Lei Xu and Dr. Yi Huang for their useful discussions and proofreading. We also acknowledge all members of ACCESS Linnaeus Centre at KTH Royal Institute of Technology, Stockholm, Sweden. Especially, we are thankful to Guodong Shi, Junfeng Wu, Kun Liu, Jie Lu, Weiguo Xia, Liquan Fu, Kin Cheong Sou, Xiangyu Meng, and Kezhi Li for their contributions and companionship during that wonderful time in Sweden. We would also like to thank our editors Professor Tamer Başar and Miss Dana Knowles for their help and professionalism. Finally, we acknowledge the support of our research by the National Natural Science Foundation of China under grants 61833009, 61873140, 61991403, and 61991400, Beijing Natural Science Foundation under grant JQ20013, the Knut and Alice Wallenberg Foundation, the Swedish Foundation for Strategic Research, and the Swedish Research Council. In addition, we acknowledge IEEE and Elsevier for granting us the permission to reuse materials from our publications copyrighted by these publishers in this book.

Beijing, China   Ziyang Meng

Shenyang, China   Tao Yang

Stockholm, Sweden   Karl H. Johansson
June 2021

Contents

Chapter 1
Introduction

Coordination of networked dynamical systems has drawn significant attention due to its broad applications in biology, social science, computer science, and engineering in recent years [14, 29, 31]. Although different problems are modeled from different disciplines, the basic assignment of these networked dynamical systems is to accomplish a common desirable objective in a cooperative manner. One approach is through a completely centralized strategy, where a single decision maker gathers the entire system's information, performs the computation, and sends back the solution to an individual agent. This centralized framework is subject to performance limitations, such as a single point of failure, high communication requirement and cost, substantial computational burden, limited flexibility and scalability, and lack of privacy. To overcome these limitations, an alternative distributed approach has received substantial attention in the area of systems and control. The idea is that by carefully designing distributed controllers, a group of autonomous agents can achieve a collective task by cooperatively and locally exchanging information. Interactions are local in the sense that an agent can only interact with a subset of agents. Under such a distributed framework, this book aims to provide some recent results on modeling, analysis, control, and applications of networked dynamical systems.

1.1 Motivating Examples

In this section, we briefly present some motivating examples of networked dynamical systems and current trends to solve the related problems.

© The Author(s), under exclusive license to Springer Nature Switzerland AG 2021
Z. Meng et al., *Modelling, Analysis, and Control of Networked Dynamical Systems*,
Systems & Control: Foundations & Applications,
https://doi.org/10.1007/978-3-030-84682-4_1

1.1.1   Smart Grid

In the past decades, the power system has been undergoing a transition from a system with conventional generation power plants and inflexible loads to a system with a large number of distributed generators, energy storages, and flexible loads, often be referred to as distributed energy resources shown in [13, 21, 23]. These resources are small and highly flexible compared with conventional generators, and can be aggregated to provide power necessary to meet the regular demand.

In order to achieve an effective deployment among distributed energy resources, one needs to properly design the coordination among them. The traditional approach is centralized, where a single control center collects all the necessary information, performs central computation, and provides control signals to the entire system. This prevalent approach has several limitations, such as a single point of failure, limited flexibility and scalability, and high communication requirement and computation burden.

Recently, an alternative distributed approach has been proposed to overcome these limitations. In particular, each unit makes a local decision based on the information received from a few neighboring units over the underlying communication network. Most existing studies on energy resource coordination focus on a single type of energy resource coordination. On the one hand, for distributed generation coordination, various distributed algorithms have been proposed. On the other hand, the problem of coordinating energy storages is challenging since the operation of storages in different time steps is interdependent due to the fact that there is only limited energy that can be stored in a storage device, see., e.g., [8, 35]. For the case of mixed types of distributed energy resources, the recent works [11, 33, 36, 37] considered the optimal distributed coordination for both distributed generators and energy storages.

1.1.2   Spacecraft Formation Flying

The study on spacecraft formation flying is a main trend of future space science. By introducing a distributed framework, many inexpensive, simple spacecraft working together can achieve the same objective as a single, expensive, and complicated spacecraft. A typical application of distributed spacecraft system is the distributed remote sensing technology [26, 27, 30]. Compared with the traditional technology, distributed remote sensing technology has the advantages of multi-source information access, global coverage, and quick revisit. Micro/nano-satellites and their distributed implementation have been shown to be well-suited for the distributed remote sensing technology because of their unique properties of mass production, low cost, and various forms. For a distributed spacecraft system, the functional implementation usually replies on its formation and attitude, and different formation and attitude guarantee different functionality to be realized. Take Darwin project as an example. This project consists of several free-flying space telescopes and a main spacecraft. Space telescopes redirect light from distant stars and planets to a main spacecraft, and the main spacecraft is responsible to communication with the earth. By effective cooperation, the formation and attitudes of these spacecraft are maintained accurately in order to accomplish the mission successfully. Another example is Planet's Dove mission, operated by Planet Labs Inc. This mission is to collect on-demand high spatial and temporal resolution optical remote sensing datum and to monitor global change by imaging [3, 6]. It consists of over 250 flock satellites, which are designed to have sufficient imaging swath width such that 100 evenly spaced nadir-pointing satellites on a single orbital plane at 500km sun synchronous obit (SSO) can image the entire earth every day at 3–5 m ground resolution. As has been indicated, one of the key problems of spacecraft formation flying is cooperative attitude control, especially for the interferometry applications, where the relative attitudes must be maintained for different spacecrafts. The general approach for networked dynamical systems can be quite useful in such a case.

1.1.3 Heavy-Duty Vehicle Platooning

Urban highways in major cities nowadays suffer from traffic congestion, which increases fuel consumption and air pollution. A recent study [24] by the International Transport Forum shows that the transport-sector carbon dioxide (CO_2) emission represents 23% globally and 30% within the Organization for Economic Cooperation and Development countries of the overall CO_2 emissions from fossil fuel combustion.

Vehicle platooning has been widely recognized as a promising solution to reduce fuel consumption and carbon dioxide emissions, to enhance the safety, and to improve highway utility, see., e.g., [2, 12, 28]. The key goal of vehicle platooning is to regulate the relative positions between neighboring vehicle in the line graph, and maintain a set velocity. Several control solutions can be traced back to the earlier work of [15, 18]. The control of vehicle platooning has attracted renewed interests due to the recent advances in manufactory vehicle industry, wireless communication, and distributed control.

1.1.4 Multiple Unmanned Aerial Vehicles

Distributed formation control of multiple unmanned aerial vehicles (UAVs) is an important application of networked dynamical systems. Instead of a single expensive and monolithic UAV, distributed formation control of multiple micro- and compact UAVs is capable of accomplishing the same maneuver efficiently without costly expenses [9]. By local information interaction, a team of UAVs cooperatively works in a distributed manner such that no member plays a central role while a disabled individual will not destroy the team function [1]. Formation of multiple UAVs has broad application prospects in large-scope searching, cluster operation, and multi-target tracking. Recently, multiple vertical takeoff and landing UAVs have been shown to be promising in military and civil fields due to their unique characteristics including hovering, vertically taking off and landing, and low-speed or low-altitude operation. A typical application project associated with multiple UAVs, called "LOCUST," has been proposed by American Office of Naval Research. This project consists of numerous inexpensive "Coyote" drones weighing no more than 13 pounds, whose endurance capacity can reach a speed of 90 mile per hour (MPH). By effective cooperation and interaction, these drones succeed in not only maintaining prescribed patterns but also enclosing a moving aircraft. To achieve the cooperative formation of multiple UAVs, distributed controls for dynamic coupling systems are critical and a number of approaches for networked dynamical systems can be useful. In addition, some practical problems, such as anti-collision, connection maintenance, and obstacle avoidance, encountered in flight maneuvers can also be addressed by extending some effective strategies associated with networked dynamical systems.

1.1.5   Social Networks

In social networks, the relationships between people are friendship, collaboration, sharing of information, and so on. In addition, there are also negative effects, including antagonistic, hostile, disagreement, conflict, and so on. Antagonistic interactions are also extensively obtained in engineering applications, e.g., in the context of target tracking applications, many objectives are noncooperative rather than cooperative. Based on these observations, the study of social networks becomes an interesting topic for networked dynamical systems. To model such an antagonistic interaction as well as cooperative interaction, a positive/negative sign is assigned indicating cooperative/antagonistic relationship. The fundamental issue is to reason about the mixture of positive and negative relationships that take place within a network. More specifically, a general model should be constructed to characterize antagonistic interactions in the network. Based on the proposed cooperative-antagonistic network model, the different behaviors are defined raised by different network structures. How to implement coordinated control algorithms and how to design information exchange structure are important to establish different behaviors. Relying on the study of behavior analysis of cooperative-antagonistic networks, fundamental properties can be revealed for social networks.

1.2   Objectives

Due to the importance of networked dynamical systems to scientific research and broad applications in various areas, this book aims to present some recent contributions on the study of networked dynamical systems, focusing on modeling, analysis, control, and applications.

From a theoretical perspective, it is valuable to model networked dynamical systems and analyze their fundamental properties of emergent behaviors. Typical emergent behaviors are observed in circuit networks, social networks, and biological networks [19, 32, 34]. Precise mathematical models are introduced to reconstruct these group behaviors including synchronizing to a time-varying trajectory, clustering into different groups, flocking with alignment, separation, and cohesion properties. Therefore, the first objective of this book is to introduce some recent results on the behavior analysis of networked dynamical systems, where synchronization, periodic behaviors, and modulus consensus are discussed, respectively.

It is essential to model networked dynamical systems in a realistic environment and to properly design their controls such that the overall objective is achieved with limited resources. For different problems, systems with different constraints are considered in this book. Typical constraints include, but are not limited to, those on dynamics, energy, communication, sensing, and control. In particular, the second objective of this book is to propose particular control algorithms for the networked models with emphasis on input saturation, large delays, and heterogenous dynamics.

From an application perspective, it is indispensable to implement the proposed distributed control algorithms into practical systems and adjust them to various engineering requirements. The typical applications include multiple unmanned aerial vehicles, spacecraft formation flying, heavy-duty vehicle platooning, smart grid, and social networks. Motivated by these facts, the first application of this book comes from spacecraft formation flying, and we intend to solve the cooperative attitude control of multiple spacecraft systems. The second application lies in multi-robot systems, and the rendezvous problem for a group of robotics is considered. The last application of this book comes from the power systems, and we target to propose distributed algorithms for a power network of a large number of distributed energy resources.

1.3   Book Structure

This book is divided into four parts. The first part presents preliminaries and the general networked dynamical model that will be repeatedly used in the rest of the book. In particular, Chap. 2 introduces notations used in this book, preliminaries on system theory, convex analysis, and definitions for interaction graphs. Then, we define the general networked dynamical model and also introduce examples of the considered models in Chap. 3.

The second part of this book focuses on the behavior analysis of networked dynamical systems. In particular, synchronization of networked dynamical systems is discussed in Chap. 4, and the extension to φ-synchronization with non-expansive dynamics is studied in Chap. 5. In addition, periodic solutions of networked dynamical systems and modulus consensus of cooperative-antagonistic networks are discussed in Chaps. 6 and 7, respectively.

The third part of this book studies the control problems of networked dynamical systems. We particularly solve the problems with input constraints in Chap. 8, large delays in Chap. 9, and heterogenous dynamics in Chap. 10, respectively.

The last part of this book gives three applications regarding the networked dynamical systems in Chaps. 11, 12, and 13, where cooperative attitude control problem of spacecraft formation flying, rendezvous problem of multi-robot systems, and energy resource coordination problem of power network are studied, respectively.

1.4 Literature

Networked dynamical systems have also been studied in other related books. For example, the author of [22] studies the cooperative control problem of dynamical systems. Cooperative stability of both linear and nonlinear systems is considered, and the theoretical results are applied to a team of unmanned ground and aerial vehicles. Reference [25] focuses on distributed coordination of multi-agent networks and introduces emergent behaviors, emergent models, and emergent issues in multi-agent systems. Then, the author of [20] emphasizes the necessity of distributed algorithms and focuses on the distributed consensus and optimization problems. The solutions of distributed averaging dynamics and gradient-based consensus algorithm are proposed. Also, distributed average consensus and weight balancing are considered in [10] under different communication topologies and constraints including directed graphs, delays, packet drops, privacy considerations, and fast convergence. Also, typical cooperative control problems are studied in [17], including consensus, distributed tracking, and containment control of systems with linear dynamics and Lipschitz nonlinear dynamics. A systematic consensus region approach is proposed to derive distributed cooperative control laws. Ref. [40] studies the solution of multi-agent networks from a distributed optimization-based control point of view. The authors of [5] present a coherent introduction to the basic distributed algorithms for robotic networks and the authors of [1] consider the coordinated motion control problem of a group of UAVs. In addition, Ref. [16] focuses on the optimal and adaptive perspectives of control of multi-agent systems, and Ref. [4] emphasizes the fundamental phenomena over networked systems, including consensus and disagreement, stable equilibria in compartmental flow networks, and synchronization in coupled oscillators. Finally, the authors of

[38] give a detailed overview of existing distributed optimization algorithms and also discuss the applications in power systems.

References

1. A. Abdessameud, A. Tayebi, *Motion Coordination for VTOL Unmanned Aerial Vehicles* (Springer, London, 2013)
2. A. Alam, A. Gattami, K.H. Johansson, An experimental study on the fuel reduction potential of heavy duty vehicle platooning, in *Proceedings of the IEEE Conference on Intelligent Transportation Systems*, Funchal, 2010, pp. 306–311
3. C.R. Boshuizen, J. Mason, P. Klupar, S. Spanhake, Results from the planet labs flock constellation, in *Proceedings of the 28th Annual AIAA/USU Conference on Small Satellites*, North Logan, 2014, pp. 1–8
4. F. Bullo, J. Cortés, F. Dörfler et al., *Lectures on Network Systems* (Kindle Direct Publishing, Santa Barbara, CA, 2019)
5. F. Bullo, J. Cortés, S. Martínez, *Distributed Control of Robotic Networks: A Mathematical Approach to Motion Coordination Algorithms* (Princeton University Press, Princeton, 2009)
6. K. Devaraj, R. Kingsbury, M. Ligon et al., Dove high speed downlink system, in *Proceedings of the 31th Annual AIAA/USU Conference on Small Satellites*, North Logan, 2017, pp. 1–8
7. D. Easley, J. Kleinberg, *Networks, Crowds, and Markets: Reasoning About a Highly Connected World* (Cambridge University Press, Cambridge, 2010)
8. H. Fang, D. Wu, T. Yang, Cooperative management of a Lithium-ion battery energy storage network: a distributed MPC approach, in *Proceedings of the IEEE 55th Conference on Decision and Control*, 2016, pp. 4226–4232
9. F. Giulietti, L. Pollini, M. Innocenti, Autonomous formation flight. IEEE Control Syst. Mag. **20**(6), 34–44 (2000)
10. C.N. Hadjicostis, A.D. Dominguez-Garcia, T. Charalambous, *Distributed Averaging and Balancing in Network Systems with Applications to Coordination and Control.* Foundations and Trends in Systems and Control, 2018
11. G. Hug, S. Kar, C. Wu, Consensus + innovations approach for distributed multiagent coordination in a microgrid. IEEE Trans. Smart Grid **6**(4), 1893–1903 (2015)
12. P.A. Ioannou, C. Chien, Autonomous intelligent cruise control. IEEE Trans. Vehic. Technol. **42**(4), 657–672 (1993)
13. R. Lasseter, A. Akhil, C. Marnay et al., Integration of distributed energy resources: the certs microgrid concept. Lawrence Berkeley National Laboratory, 2002
14. N.E. Leonard, D.A. Paley, F. Lekien et al., Collective motion, sensor networks, and ocean sampling. Proc. IEEE **95**(1), 48–74 (2007)
15. W.S. Levine, M. Athans, On the optimal error regulation of a string of moving vehicles. IEEE Trans. Autom. Control **11**(3), 355–361 (1966)
16. F.L. Lewis, H. Zhang, K. Hengster-Movric et al., *Cooperative Control of Multi-agent Systems: Optimal and Adaptive Design Approaches* (Springer, London, 2014)
17. Z. Li, Z. Duan, *Cooperative Control of Multi-Agent Systems: A Consensus Region Approach* (CRC Press, Boca Raton, 2017)
18. S.M. Melzer, B.C. Kuo, Optimal regulation of systems described by a countably infinite number of objects. Automatica **7**(3), 359–366 (1971)
19. M. Nagy, Z. Akos, D. Biro et al., Hierarchical group dynamics in pigeon flocks. Nature **464**(7290), 890–893 (2010)
20. A. Nedić, *Convergence Rate of Distributed Averaging Dynamics and Optimization in Networks.* Foundations and Trends in Systems and Control (Hanover, Delft, 2015)

21. M.A.A. Pedrasa, T.D. Spooner, I.F. MacGill, Coordinated scheduling of residential distributed energy resources to optimize smart home energy services. IEEE Trans. Smart Grid **1**(2), 134–143 (2010)
22. Z. Qu, *Cooperative Control of Dynamical Systems: Applications to Autonomous Vehicles* (Springer, London, 2009)
23. F. Rahimi, A. Ipakchi, Demand response as a market resource under the smart grid paradigm. IEEE Trans. Smart Grid **1**(1), 82–88 (2010)
24. Reducing transport greenhouse gas emission: Trends & data. OECD/ITF, 2010
25. W. Ren, Y. Cao, *Distributed Coordination of Multi-agent Networks: Emergent Problems, Models, and Issues* (Springer, London, 2011)
26. R.S. Smith, F.Y. Hadaegh, Control of deep-space formation flying spacecraft: relative sensing and switched information. J. Guidance Control Dynam. **28**(1), 106–114 (2005)
27. M.C. VanDyake, C.D. Hall, Decentralized coordinated attitude control of a formation of spacecraft. J. Guidance Control Dynam. **29**(5), 1101–1109 (2006)
28. P. Varaiya, Smart cars on smart roads: problems of control. IEEE Trans. Autom. Control **38**(2), 195–207 (1993)
29. A. Vespignani, Predicting the behavior of techno-social systems. Science **325**(5939), 425–428 (2009)
30. P.K.C. Wang, F.Y. Hadaegh, Coordination and control of multiple microspacecraft moving in formation. J. Astron. Sci. **44**(3), 315–355 (1996)
31. D.J. Watts, S.H. Strogatz, Collective dynamics of small-world networks. Nature **393**(6648), 440–442 (1998)
32. C.W. Wu, L.O. Chua, Synchronization in an array of linearly coupled dynamical systems. IEEE Trans. Circ. Syst. I Fund. Theory Appl. **42**(8), 430–447 (1995)
33. D. Wu, T. Yang, A.A. Stoorvogel et al., Distributed optimal coordination for distributed energy resources in power systems. IEEE Trans. Autom. Sci. Eng. **14**(2), 414–424 (2017)
34. W. Xia, M. Cao, Clustering in diffusively coupled networks. Automatica **47**(11), 2395–2405 (2011)
35. Y. Xu, W. Zhang, G. Hug et al., Cooperative control of distributed energy storage systems in a microgrid. IEEE Trans. Smart Grid **6**(1), 238–248 (2015)
36. T. Yang, D. Wu, W. Ren et al., Cooperative optimal coordination for distributed energy resources, in *Proceedings of the IEEE Conference on Decision and Control*, Melbourne, 2017, pp. 6334–6339
37. T. Yang, D. Wu, A.A. Stoorvogel et al., Distributed coordination of energy storage with distributed generators, in *Proceedings of the IEEE Power and Energy Society General Meeting*, Boston, 2016
38. T. Yang, X. Yi, J. Wu et al., A survey of distributed optimization. Annu. Rev. Control **47**, 278–305 (2019)
39. W. Zachary, An information flow model for conflict and fission in small groups. J. Anthropol. Res. **33**(4), 452–473 (1977)
40. M. Zhu, S. Martínez, *Distributed Optimization-based Control of Multi-Agent Networks in Complex Environments* (Springer International Publishing, New York, 2015)

Part I
Modelling

Chapter 2
Preliminaries

In this chapter, we introduce notations used in this book, system theory, convex analysis, and basic definitions for interaction graphs.

Notations

$\equiv$	identically equal
$\triangleq$	define as
$\forall$	for all
$\exists$	if there exists
$\Rightarrow$	implies
$\in$	belongs to
$\notin$	does not belongs to
$\subset$	a strict subset of
$\subseteq$	a subset of
$\cup$	union
$\cap$	intersection
$\backslash$	excludes
$\mathbb{R}$	the set of real numbers
$\mathbb{R}^n$	the set of $n \times 1$ real vectors
$\mathbb{R}^{n \times m}$	the set of $n \times m$ real matrices
$\mathbb{Z}$	the set of integers
$\mathbb{R}_+$	the set of non-negative real numbers
$\mathbb{Z}_+$	the set of non-negative integers
$\mathbb{C}$	the set of complex numbers
$B(x, \epsilon)$	the open ball centered at x with radius ϵ
$\text{Re}(z)$	the real part of a complex number z
$\text{Im}(z)$	the imaginary part of a complex number z

© The Author(s), under exclusive license to Springer Nature Switzerland AG 2021
Z. Meng et al., *Modelling, Analysis, and Control of Networked Dynamical Systems*,
Systems & Control: Foundations & Applications,
https://doi.org/10.1007/978-3-030-84682-4_2

$\lfloor c \rfloor$	the largest integer less than or equal to c for a given real number c
$\sum$	summation
max	maximum
min	minimum
sup	supremum, the least upper bound
inf	infimum, the greatest lower bound
∞	infinity
A^{T}	the transpose of A
rank(A)	rank of matrix A
$\|A\|$	the induced norm of a matrix A
I_n	the $n \times n$ identity matrix
$\mathbf{1}_n$	the $n \times 1$ vector of all ones
$\mathbf{0}_n$	the $n \times 1$ vector of all zeros
$S(\cdot)$	a 3×3 skew-symmetric matrix for a 3×1 vector
$A > 0$	a positive definite matrix A
$A \geq 0$	a positive semi-definite matrix A
$A^{\frac{1}{2}}$ or $\sqrt{A}$	the square root of a positive definite matrix A
$\lambda_{\min}(A)$	the maximum eigenvalue of a real symmetric matrix A
$\lambda_{\max}(A)$	the minimum eigenvalue of a real symmetric matrix A
$\mathrm{sgn}(x_1)$	the sign function of $x_1 \in \mathbb{R}$, i.e., $\mathrm{sgn}(x_1) = \begin{cases} 1, & x_1 > 0 \\ 0, & x_1 = 0 \\ -1 & x_1 < 0 \end{cases}$
$\mathrm{sgn}(x)$	$\mathrm{sgn}(x) = [\mathrm{sgn}(x_1), \ldots, \mathrm{sgn}(x_n)]^{\mathrm{T}}$ for $x = [x_1 \ldots, x_n]^{\mathrm{T}} \in \mathbb{R}^n$
$\|x\|$	Euclidean norm of a vector $x \in \mathbb{R}^n$
$\|x\|_1$	1-norm of a vector $x \in \mathbb{R}^n$
$\|x\|_P$	the Hilbert norm of $x \in \mathbb{R}^n$ with $P \in \mathbb{R}^{n \times n}$ being a symmetric positive definite matrix
$|x|$	$|x| = [|x_1|, |x_2|, \ldots, |x_n|]^{\mathrm{T}}$, $x = [x_1, x_1, \ldots, x_n]^{\mathrm{T}} \in \mathbb{R}^n$
diag$(a_1, \ldots, a_n)$	a diagonal matrix with diagonal entries a_1 to a_n
diag$(A_1, \ldots, A_n)$	a block diagonal matrix with diagonal blocks A_1 to A_n
$A \otimes B$	the Kronecker product between matrices A and B
$a < (\leq)b$	each entry of $a - b$ is negative (non-positive), for $a, b \in \mathbb{R}^n$
$a > (\geq)b$	each entry of $a - b$ is positive (non-negative), for $a, b \in \mathbb{R}^n$
$\mathbf{j}$	$\mathbf{j} = \sqrt{-1}$
$\overline{w}$	the complex conjugate of a complex number $w \in \mathbb{C}$
sin	sine function
cos	cosine function
$\mathcal{G}$	$\mathcal{G} = (\mathcal{V}, \mathcal{E})$ denotes graph consisting set of agents and set of edges
$\mathcal{V}$	agent set $\{1, 2, \ldots, N\}$ with N being a positive constant
$\mathcal{E}$	$\mathcal{E} \subseteq \mathcal{V} \times \mathcal{V}$ denotes set of edges
$\mathcal{N}_i$	set of neighbors for agent i

2.1 System Theory

We recall the following definitions and lemmas from linear and nonlinear system theory.

Definition 2.1

- A symmetric matrix A is positive (negative) definite if and only if all its eigenvalues are positive (negative).
- A symmetric matrix A is positive (negative) semi-definite if and only if all its eigenvalues are non-negative (non-positive).
- A square matrix A is called a Hurwitz matrix if every eigenvalue of A has strictly negative real part.

Definition 2.2 Consider a linear time-invariant (LTI) system $\dot{x} = Ax$, where $x \in \mathbb{R}^n$ and $A \in \mathbb{R}^{n \times n}$, the LTI system is

- Asymptotically stable if and only if all the eigenvalues of A are in the open left half complex plane.
- Neutrally stable if and only if all the eigenvalues of A are in the closed left half complex plane, and those on $j\omega$-axis are simple.
- Unstable if and only if there is at least one eigenvalue of A in the open right half complex plane.

The above definitions will be used in Chaps. 4, 5, 6, 8, 9, and 10.

Definition 2.3 A function $f(t, x) : [0, \infty) \times \mathbb{R}^n \rightarrow \mathbb{R}^n$ is globally Lipschitz continuous in $x \in \mathbb{R}^n$ if there exists a constant $\delta > 0$ such that

$$\| f(t, x) - f(t, y) \| \le \delta \| x - y \|, \forall x, y \in \mathbb{R}^n, \forall t \ge 0. \tag{2.1}$$

The above definitions will be used in Chaps. 4 and 12.

2.1.1 LaSalle's Invariance Principle

Definition 2.4 Given the system

$$\dot{x} = f(x), \tag{2.2}$$

where $f : \mathcal{D} \rightarrow \mathbb{R}^n$ is a locally Lipschitz function from a domain $\mathcal{D} \subset \mathbb{R}^n$ into $\mathbb{R}^n$. Let $x(t)$ be a solution of (2.2). A set $\mathcal{M}$ is said to a positively invariant set if $x(0) \in \mathcal{M} \Rightarrow x(t) \in \mathcal{M}, \forall t \ge 0$.

Lemma 2.5 (LaSalle's Invariance Principle) *Let $\Omega \subset \mathcal{D}$ be a compact set that is positively invariant with respect to (2.2). Let $V : \mathcal{D} \rightarrow \mathbb{R}$ be continuously*

differentiable function such that $\dot{V}(x) \leq 0$ *in* Ω. *Let* M *be the set of all points in* Ω
where $\dot{V}(x) = 0$. *Then every solution starting in* Ω *approaches* M *as* $t \to \infty$.

The above lemma will be used in Chap. 12.

2.1.2 Dini Derivatives

Let $D^+V(t, x(t))$ be the upper Dini derivative of $V(t, x(t))$ with respect to t, *i.e.*,

$$D^+V(t, x) = \lim_{\varepsilon \to 0^+} \sup \frac{V(t + \varepsilon, x(t + \varepsilon)) - V(t, x(t))}{\varepsilon}.$$

The following lemma holds.

Lemma 2.6 *Consider* $V = \{1, 2, \ldots, N\}$ *is a finite and nonempty set. Suppose for each* $i \in V$, $V_i : \mathbb{R} \times \mathbb{R}^n \to \mathbb{R}$ *is continuously differentiable. Let* $V(t, x) = \max_{i \in V} V_i(t, x)$, *and let* $\overline{V}(t) = \{i \in V : V_i(t, x(t)) = V(t, x(t))\}$ *be the set of indices where the maximum is reached at time* t. *Then*

$$D^+V(t, x(t)) = \max_{i \in \overline{V}(t)} \dot{V}_i(t, x(t)).$$

The above definitions and lemmas will be used in Chaps. 4, 5, and 9.

2.2 Convex Analysis

For any nonempty set $S \subseteq \mathbb{R}^n$, we use $d(x, S) = \inf_{y \in S} \|x - y\|$ to denote the distance between $x \in \mathbb{R}^n$ and S. Obviously, $d(x, S) = 0$, for $x \in S$. A set $S \subset \mathbb{R}^n$ is said to be convex if $(1 - \zeta)x + \zeta y \in S$ when $x \in S$, $y \in S$, and $0 \leq \zeta \leq 1$.

Let S be a convex set. The convex projection of any $x \in \mathbb{R}^n$ onto S is denoted by $P_S(x) \in S$ satisfying $\|x - P_S(x)\| = d(x, S)$. We also know that $d^2(x, S)$ is continuously differentiable for all $x \in \mathbb{R}^m$, and its gradient can be explicitly obtained by:

$$\nabla d^2(x, S) = 2(x - P_S(x)). \tag{2.3}$$

Also, it is easy to see that

$$(P_S(x) - x)^T(P_S(x) - y) \leq 0, \quad \forall y \in S. \tag{2.4}$$

In addition,

$$\|P_S(x) - P_S(y)\| \le \|x - y\| \quad \forall x, y \in \mathbb{R}^n. \tag{2.5}$$

We next give the definition on convex function.

Definition 2.7

- A function $f : \mathbb{R}^n \to \mathbb{R}$ is convex if

$$f(\theta x + (1 - \theta)y) \le \theta f(x) + (1 - \theta)f(y), \forall x, y \in \mathbb{R}^n, \forall 0 \le \theta \le 1. \tag{2.6}$$

- A function $f : \mathbb{R}^n \to \mathbb{R}$ is strictly convex if strict inequality holds in (2.6) whenever $x \ne y$ and $0 < \theta < 1$.
- A function $f : \mathbb{R}^n \to \mathbb{R}$ is strongly convex if there exists an $m > 0$ such that

$$f(y) \ge f(x) + \nabla f^\mathsf{T}(x)(y - x) + \frac{m}{2}\|y - x\|_2^2. \tag{2.7}$$

The above definitions and properties will be used in Chaps. 5, 12, and 13.

2.3 Interaction Graphs

We recall some basic concepts from graph theory to characterize the information exchange in the networked dynamical systems. A directed or undirected graph/communication topology $\mathcal{G}$ consists of a pair $(\mathcal{V}, \mathcal{E})$, where $\mathcal{V} = \{1, 2, \ldots, N\}$ is a finite, nonempty set of nodes and $\mathcal{E} \subseteq \mathcal{V} \times \mathcal{V}$ is a set of ordered pairs of nodes. An edge/arc (i, j) denotes that nodes j can obtain information from i. An undirected graph is defined such that $(j, i) \in \mathcal{E}$ implies $(i, j) \in \mathcal{E}$. In such a case, an edge $\{j, i\} \in \mathcal{E}$ denotes that node i, j can obtain each other's information mutually. All the neighbors of node i are denoted as $\mathcal{N}_i := \{j | (j, i) \in \mathcal{E}\}$, where we assume that $i \notin \mathcal{N}_i$.

For a graph $\mathcal{G}$, its adjacency matrix $A = [a_{ij}] \in \mathbb{R}^{N \times N}$ is defined such that a_{ij} is positive if $(j, i) \in \mathcal{E}$ and $a_{ij} = 0$ otherwise. Here we assume that $a_{ii} = 0$, for all $i \in \mathcal{V}$. The Laplacian matrix $L = [l_{ij}] \in \mathbb{R}^{N \times N}$ associated with A is defined as $l_{ii} = \sum_{j \ne i} a_{ij}$ and $l_{ij} = -a_{ij}$, where $i \ne j$.

A directed path in a directed graph or an undirected path in an undirected graph is a sequence of edges of the form $(i_1, i_2), (i_2, i_3), \ldots$. If there exists a path from node i to j, then node j is said to be reachable from node i. $\mathcal{G}$ is said to be strongly connected if each node is reachable from any other node. A directed graph is quasi-strongly connected if it has a directed spanning tree, i.e., there exists at least one node that is reachable to all other nodes. An undirected graph is connected if it is connected as an undirected graph ignoring the edge directions.

To this end, we have given a clear definition for the graph theory with a fixed communication topology. On the other hand, one common issue of equipping communication unit for networked dynamical systems is the possible communication link failure. Therefore, it is necessary to consider the validness of the proposed algorithms for the case of switching communication topology. We therefore associate the switching communication topology with a time-varying graph $\mathcal{G}_{\sigma(t)} = (\mathcal{V}, \mathcal{E}_{\sigma(t)})$, where $\sigma : [0, +\infty) \to \mathcal{P}$ is a piecewise constant function and $\mathcal{P}$ is finite set of all possible graphs. $\mathcal{G}_{\sigma(t)}$ remains constant for $t \in [t_\ell, t_{\ell+1})$, $\ell = 0, 1, \ldots$ and switches at $t = t_\ell$, $\ell = 1, \ldots$. In addition, we assume that $\inf_\ell(t_{\ell+1} - t_\ell) \geq \tau_D > 0$ for the continuous-time setting, $\ell = 1, \ldots$, where τ_D is a constant known as dwell time. The joint graph of $\mathcal{G}_{\sigma(t)}$ during time interval $[t_1, t_2)$ is defined by $\mathcal{G}_{\sigma(t)}([t_1, t_2)) = \bigcup_{t \in [t_1, t_2)} \mathcal{G}(t) = (\mathcal{V}, \bigcup_{t \in [t_1, t_2)} \mathcal{E}(t))$. Moreover, j is a neighbor of i at time t when $(j, i) \in \mathcal{E}_{\sigma(t)}$, and $\mathcal{N}_i(\sigma(t))$ represents the set of agent i's neighbors at time t.

In the following, we will give definitions on the joint connectivity.

Definition 2.8 $\mathcal{G}_{\sigma(t)}$ is *uniformly jointly strongly connected* if there exists a constant $T > 0$ such that $\mathcal{G}([t, t+T))$ is strongly connected for any $t \geq 0$.

Definition 2.9 $\mathcal{G}_{\sigma(t)}$ is *uniformly jointly quasi-strongly connected* if there exists a constant $T > 0$ such that $\mathcal{G}([t, t+T))$ is quasi-strongly connected for any $t \geq 0$.

Definition 2.10 Assume that $\mathcal{G}_{\sigma(t)}$ is undirected for all $t \geq 0$. $\mathcal{G}_{\sigma(t)}$ is *uniformly jointly connected* if there exists a constant $T > 0$ such that $\mathcal{G}([t, t+T))$ is connected for any $t \geq 0$.

Definition 2.11 Assume that $\mathcal{G}_{\sigma(t)}$ is undirected for all $t \geq 0$. $\mathcal{G}_{\sigma(t)}$ is *infinitely jointly connected* if $\mathcal{G}([t, \infty))$ is connected for any $t \geq 0$.

We next present several examples to illustrate the differences among different connectivity definitions. In the first place, we consider that the communication topology switches periodically as Fig. 2.1 when the systems are at time instants $t_\ell = \ell s$, $\ell = 0, 1, \ldots$ and remain constant during $[t_{\ell-1}, t_\ell)$, $\ell = 1, 2, \ldots$. The graphs $\mathcal{G}_1, \mathcal{G}_2, \mathcal{G}_3$ are represented in Figs. 2.2, 2.3, and 2.4. It is not hard to show that $\mathcal{G}([t, t+3))$ is strongly connected for any $t \geq 0$ in such a case and therefore Definition 2.8 is satisfied. In the second place, we consider that the communication topology switches periodically as Fig. 2.5 when the systems are at time instants $t_\ell = \ell s$, $\ell = 0, 1, \ldots$ and remain constant during $[t_{\ell-1}, t_\ell)$, $\ell = 1, 2, \ldots$. The graphs $\mathcal{G}_1$ and $\mathcal{G}_2$ are represented in Figs. 2.2 and 2.3. It is not hard to show that

Fig. 2.1 Switching communication topology $\mathcal{G}_1 \longrightarrow \mathcal{G}_2 \longrightarrow \mathcal{G}_3 \longrightarrow \mathcal{G}_1 \longrightarrow \cdots$

Fig. 2.2 $\mathcal{G}_1$

Fig. 2.3 $\mathcal{G}_2$

Fig. 2.4 $\mathcal{G}_3$

Fig. 2.5 Switching
communication topology

$\mathcal{G}([t, t + 2))$ is quasi-strongly connected for any $t \geq 0$ in such a case and therefore Definition 2.9 is satisfied. In the third place, we still consider that the communication topology switches periodically as Fig. 2.1 when the systems are at time instants $t_\ell = \ell\, s$, $\ell = 0, 1, \ldots$ and remain constant during $[t_{\ell-1}, t_\ell)$, $\ell = 1, 2, \ldots$. The graphs $\mathcal{G}_1$, $\mathcal{G}_2$, $\mathcal{G}_3$ are represented in Figs. 2.6, 2.7, and 2.8 in this case. It is clear that $\mathcal{G}([t, t+3))$ is connected for any $t \geq 0$ and therefore Definition 2.10 is satisfied. Last but not the least, we consider the communication topology switches between $\mathcal{G}_1$ and $\mathcal{G}_2$ in Figs. 2.9 and 2.10. The graph remains $\mathcal{G}_1$ while during time intervals $[\ell^2, \ell^2 + 1]$, the graph is $\mathcal{G}_2$, $\ell = 1, 2, \ldots$. It is not hard to show that $\mathcal{G}([t, t + \infty))$ is connected for any $t \geq 0$ in such a case and therefore Definition 2.11 is satisfied. Also note that we cannot find a constant T such that Definition 2.10 is satisfied.

The definitions in this section will be used in all the following chapters.

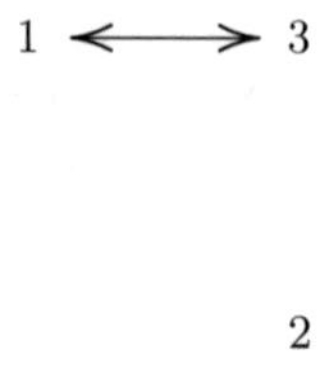

$$1 \longleftrightarrow 3$$

2

Fig. 2.6 $\mathcal{G}_1$

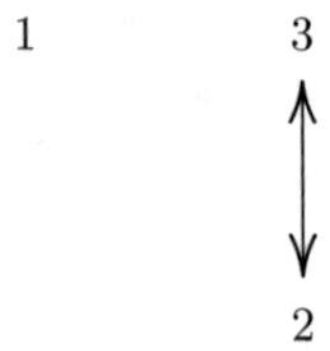

1   3

2

Fig. 2.7 $\mathcal{G}_2$

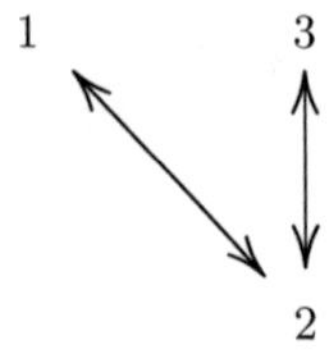

1   3

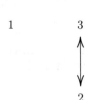

2

Fig. 2.8 $\mathcal{G}_3$

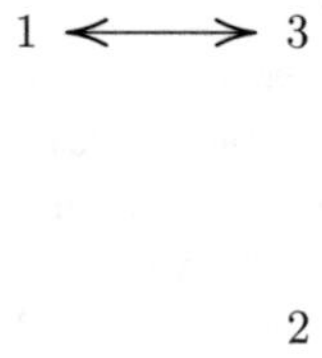

$$1 \longleftrightarrow 3$$

2

Fig. 2.9 $\mathcal{G}_1$

1   3

2

Fig. 2.10 $\mathcal{G}_2$

2.4 Literature

Lemma 2.5, i.e., LaSalle's invariance principle, is given in [5]. Equation (2.3), i.e., the gradient of a continuously differentiable function is obtained by [1]. Definition 2.7, i.e., the definition on convex function, comes from [2]. Lemma 2.6, i.e., a lemma for Dini derivative is taken from [3, 7]. In addition, the definition on the generalized time derivative of a locally Lipschitz function is given in [9, 10]. Finally, graph theory comes from [4, 8] and the definition on dwell time is originated from [6].

References

1. J.P. Aubin, *Viability Theory* (Birkhauser Boton, Boston, 1991)
2. S. Boyd, L. Vandenberghe, *Convex Optimization* (Cambridge University Press, Los Angeles, 2004)
3. J.M. Danskin, The theory of max-min, with applications. SIAM J. Appl. Math. **14**(4), 641–664 (1966)
4. C. Godsi, G.F. Royle, *Algebraic Graph Theory* (Springer, New York, 2001)
5. H.K. Khalil, *Nonlinear Systems*, 3rd edn. (Prentice-Hall, Upper Saddle River, 2002)
6. D. Liberzon, A.S. Morse, Basic problems in stability and design of switched systems. IEEE Control Syst. Mag. **19**(5), 59–70 (1999)
7. Z. Lin, B. Francis, M. Maggiore, State agreement for continuous-time coupled nonlinear systems. SIAM J. Control Optim. **46**(1), 288–307 (2007)
8. B. Mohar, Y. Alavi, The Laplacian spectrum of graphs. Graph Theory Combin. Appl. **2**, 871–898 (1991)
9. B.E. Paden, S.S. Sastry, A calculus for computing filippov's differential inclusion with application to the variable structure control of robot manipulators. IEEE Trans. Circ. Syst. I: Regul. Pap. **34**(1), 73–82 (1987)
10. D. Shevitz, B.E. Paden, Lyapunov stability theory of nonsmooth systems. IEEE Trans. Autom. Control **39**(9), 1910–1914 (1994)

Chapter 3
Networked Dynamical System Models

In this chapter, we first introduce several commonly used networked dynamical models. We then introduce examples of the considered networked dynamical models.

3.1 Continuous-Time Model

Consider a networked dynamical system consisting of N agents. Each agent has its own dynamics and the local interaction information is used to accomplish overall objective. In addition, as mentioned in Chap. 2, the information exchange may be switching and the switching communication topology is characterized by a time-varying graph $\mathcal{G}_{\sigma(t)} = (\mathcal{V}, \mathcal{E}_{\sigma(t)})$. For example, $\mathcal{G}_{\sigma(t)}$ switches among three fixed topologies $\mathcal{G}_1, \mathcal{G}_2$, and $\mathcal{G}_3$ shown in Fig. 3.1 periodically.

The state of an agent might represent physical variables including opinion, frequency, attitude, position, temperature, and voltage. The continuous-time dynamics of each agent is modeled as

$$\dot{x}_i = f_i(t, x_i, u_i),$$
$$y_i = g_i(t, x_i, u_i), \quad i \in \mathcal{V}, \tag{3.1}$$

where $x_i \in \mathbb{R}^{n_i}$ is the state of agent i, f_i and g_i are functions, $u_i \in \mathbb{R}^{m_i}$ is the control input, and y_i is the output. The available information for each agent is the information from its neighbors, and the control input is of form

$$u_i = u_i(x_i, \{x_j\}_{j \in \mathcal{N}_i(\sigma(t))}), \quad i \in \mathcal{V}.$$

© The Author(s), under exclusive license to Springer Nature Switzerland AG 2021
Z. Meng et al., *Modelling, Analysis, and Control of Networked Dynamical Systems*,
Systems & Control: Foundations & Applications,
https://doi.org/10.1007/978-3-030-84682-4_3

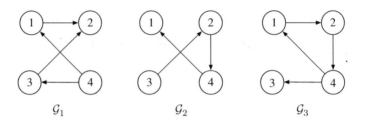

Fig. 3.1 Time-varying directed communication topology

3.2 Discrete-Time Model

Other than continuous-time model, the networked dynamical system with a discrete-time model is also considered in this book. For the discrete-time model, the switching interaction graph of the network is defined as a sequence of graphs, $\mathcal{G}_k = (\mathcal{V}, \mathcal{E}_k)$, $k = 0, 1, \ldots$, where $\mathcal{V}$ is the node set and $\mathcal{E}_k \subseteq \mathcal{V} \times \mathcal{V}$ is the set of edges at time k. An edge from node i to j is denoted as (i, j). The set of in-neighbors of node i in $\mathcal{G}_k$ is denoted as $\mathcal{N}_i^{in}(k) := \{j : (j, i) \in \mathcal{E}_k\}$. We also abuse the notation to use $\mathcal{N}_i(k)$ as $\mathcal{N}_i^{in}(k)$. The set of out-neighbors of node i in $\mathcal{G}_k$ is denoted as $\mathcal{N}_i^{out}(k) := \{j : (i, j) \in \mathcal{E}_k\}$. The joint graph of $\mathcal{G}_k$ during time interval $[k_1, k_2)$ is defined by $\mathcal{G}([k_1, k_2)) = \bigcup_{k \in [k_1, k_2)} \mathcal{G}(k) = (\mathcal{V}, \bigcup_{k \in [k_1, k_2)} \mathcal{E}_k)$. The cardinality of $\mathcal{N}_i^{out}(k)$ is called its out-degree at time k and is denoted by $d_i(k) = |\mathcal{N}_i^{out}(k)|$. We introduce the following definition on the joint connectivity of a sequence of graphs.

Definition 3.1

(i) $\{\mathcal{G}_k\}_0^\infty$ is uniformly jointly strongly connected if there exists a constant $T \geq 1$ such that $\mathcal{G}([k, k + T))$ is strongly connected for any $k \geq 0$.

(ii) Suppose $\mathcal{G}_k$ is undirected for all $k \geq 0$. Then $\{\mathcal{G}_k\}_0^\infty$ is infinitely jointly connected if $\mathcal{G}([k, +\infty))$ is connected for any $k \geq 0$.

The discrete-time dynamics of each agent is modeled as

$$x_i(k + 1) = f_i(k, x_i(k), u_i(k)),$$
$$y_i(k) = g_i(k, x_i(k), u_i(k)), \quad i \in \mathcal{V}, \quad \forall k = 0, 1, \ldots, \qquad (3.2)$$

where $x_i(k) \in \mathbb{R}^{n_i}$ is the state of node i, f_i and g_i are functions, $u_i(k) \in \mathbb{R}^{m_i}$ is the control input, and $y_i(k)$ is the output. The available information for each agent is the information from its neighbors and the control input is of form

$$u_i(k) = u_i(x_i, \{x_j\}_{j \in \mathcal{N}_i(k)}), \quad i \in \mathcal{V}, \quad \forall k = 0, 1, \ldots.$$

3.3 Leader–Follower Model

We also distinguish leaderless network and leader–follower network in this book. In particular, $\mathcal{G}$ is used to represent a leaderless network. For a leader–follower network, besides follower agents 1 to N, there is also a leader, labeled as agent 0. A leader–follower graph is defined as $\bar{\mathcal{G}} := (\bar{\mathcal{V}}, \bar{\mathcal{E}})$ with $\bar{\mathcal{V}} = \mathcal{V} \cup \{0\}$ and $\bar{\mathcal{E}} \subseteq \bar{\mathcal{V}} \times \bar{\mathcal{V}}$. We abuse the notation to use $\mathcal{G} := (\mathcal{V}, \mathcal{E})$ as the follower graph of a leader–follower network. The interaction from the leader to the follower is always undirected and the interaction weight is defined such that, for each $i = 1, 2, \ldots, N$, a_{i0} is positive if $(0, i) \in \bar{\mathcal{E}}$ and $a_{i0} = 0$ otherwise. We can easily define the adjacency matrix $\bar{A} = [a_{ij}] \in \mathbb{R}^{(N+1)\times(N+1)}$ for a leader–follower network. Then, the grounded Laplacian matrix $\bar{L} = [l_{ij}] \in \mathbb{R}^{N\times N}$ of a leader–follower graph $\bar{\mathcal{G}}$ is defined as $\bar{L} = L + \text{diag}(a_{10}, a_{20}, \ldots, a_{N0})$, where L is the Laplacian matrix associated with the follower graph $\mathcal{G}$.

Definition 3.2 $\bar{\mathcal{G}}$ is *leader rooted* if for any follower agent $i \in \mathcal{V}$ there is a directed path from the leader 0 to follower agent i.

Figure 3.2 is an example to illustrate a leader rooted graph.

Lemma 3.3

- *For a general leader–follower graph, all the eigenvalues of $\bar{L}$ are in the closed right-half plane and those on the imaginary axis are simple.*
- *For a leader rooted graph, all the eigenvalues of $\bar{L}$ are in the open right-half plane.*

A switching leader–follower communication topology is defined as $\bar{\mathcal{G}}_{\sigma(t)}$ and switches from a finite set $\{\bar{\mathcal{G}}_p\}_{p\in\mathcal{P}}$ by abusing the notation, where $\mathcal{P}$ is a finite set of all possible graphs. In addition, $\bar{\mathcal{N}}_i(\sigma(t))$ represents the set of agent i's neighbors at time t. Let the sets $\{\bar{A}_p\}_{p\in\mathcal{P}}$ and $\{\bar{L}_p\}_{p\in\mathcal{P}}$ be the leader–follower adjacency matrices and leader–follower grounded Laplacian matrices associated with $\{\bar{\mathcal{G}}_p\}_{p\in\mathcal{P}}$, respectively. Consequently, the time-varying leader–follower adjacency matrix and time-varying leader–follower grounded Laplacian matrix are defined as $\bar{A}_{\sigma(t)} = [a_{ij}(\sigma)] \in \mathbb{R}^{(N+1)\times(N+1)}$ and $\bar{L}_{\sigma(t)} = [l_{ij}(\sigma)] \in \mathbb{R}^{N\times N}$.

For the leader–follower communication topology, we introduce the following definition for switching graphs.

Fig. 3.2 Information flow associated with one leader and six followers

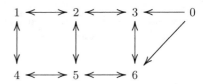

Definition 3.4

(i) $\bar{\mathcal{G}}_{\sigma(t)}$ is *jointly leader connected* in the time interval $[t_1, t_2)$ if the union graph $\bar{\mathcal{G}}([t_1, t_2))$ is leader connected.

(ii) $\bar{\mathcal{G}}_{\sigma(t)}$ is *uniformly jointly leader connected* if there exists a constant $T > 0$ such that the union graph $\bar{\mathcal{G}}([t, t + T))$ is leader connected for any $t \geq 0$.

(iii) Assume that the follower graph $\mathcal{G}_{\sigma(t)}$ is undirected for all $t \geq 0$. $\bar{\mathcal{G}}_{\sigma(t)}$ is *infinitely jointly leader connected* if the union graph $\bar{\mathcal{G}}([t, \infty))$ is leader connected for any $t \geq 0$.

For agent dynamics, the follower agents are subject to those of (3.1) while the leader agent is subject to

$$\dot{x}_0 = f_0(t, x_0),$$

$$y_0 = g_0(t, x_0), \tag{3.3}$$

where $x_0 \in \mathbb{R}^{n_0}$ is the state of node i, f_0 and g_0 are functions, and y_0 is the output. The available information for each follower agent is the information from its neighbors, and the control input is of form

$$u_i = u_i(x_i, \{x_j\}_{j \in \bar{N}_i(\sigma(t))}), \quad i \in \mathcal{V}.$$

3.4   Examples

We next introduce examples of the considered networked dynamical system model. In particular, cooperative attitude tracking of spacecraft formation flying, the rendezvous of a group of robots, and formation control of multiple unmanned aerial vehicles are illustrated.

3.4.1   Spacecraft Formation Flying

Spacecraft formation flying is an important application of networked dynamical systems. See Fig. 3.3 for illustration. In addition, cooperative attitude tracking of the spacecraft formation flying is an indispensable and difficult problem being studied for decades. We specify the dynamics of (3.1) as those of spacecraft attitude:

$$\dot{\sigma}_i = \mathbf{G}_i(\sigma_i)\omega_i,$$

$$J_i\dot{\omega}_i = -\omega_i \times (J_i\omega_i) + \tau_i, \quad i \in \mathcal{V},$$

Fig. 3.3 Spacecraft
formation flying

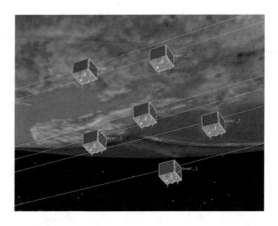

where $\sigma_i \in \mathbb{R}^3$ is the modified Rodriguez parameters (MRPs) denoting the rotation
from the body frame of the ith spacecraft to the inertial frame, and ω_i is the angular
velocity of the ith spacecraft. Also, $\mathbf{G}_i(\sigma_i)$ is given by

$$\mathbf{G}_i(\sigma_i) = \frac{1}{2}\left(S(\sigma_i) + \sigma_i\sigma_i^{\mathrm{T}} + \frac{1 - ||\sigma_i||^2}{2}I_3\right),$$

where $S(\cdot)$ denotes a 3×3 skew-symmetric matrix for a 3×1 vector, $||\sigma_i|| = \sqrt{\sigma_i^{\mathrm{T}}\sigma_i}$, and $J_i \in \mathbb{R}^{3\times3}$ and $\tau_i \in \mathbb{R}^3$ are, respectively, the inertia tensor and
control torque of the ith spacecraft. The objective is to drive the states of all the
spacecraft to the desired ones. In particular, cooperative attitude tracking is achieved
if $\lim_{t\to\infty}(\sigma_i(t) - \sigma_d(t)) = 0$ and $\lim_{t\to\infty}(\omega_i(t) - \omega_d(t)) = 0$ for all $i \in \mathcal{V}$, where
$\sigma_d(t)$ and $\omega_d(t)$ denote the desired rotation and angular velocity, respectively.

3.4.2 Multi-Robot Systems

The study on multi-robot systems attracts much attention recently due to its potential
applications in environment monitoring, search and rescue, and entertainment (see
Fig. 3.4 for illustration). The rendezvous problem for a group of robots is an
important problem for the applications of multi-robot systems. We specify the
dynamics of (3.1) as the Lagrangian equation:

$$M_i(q_i)\ddot{q}_i + C_i(q_i, \dot{q}_i)\dot{q}_i = \tau_i, \tag{3.5}$$

where $q_i \in \mathbb{R}^n$ is the vector of generalized coordinates, $M_i(q_i) \in \mathbb{R}^{n\times n}$ is the
$m \times m$ inertia (symmetric) matrix, $C_i(q_i, \dot{q}_i)\dot{q}_i$ is the Coriolis and centrifugal terms,
and $\tau_i \in \mathbb{R}^n$ is the control force. Besides the dynamics of each robotic, the individual

Fig. 3.4 Multi-robot systems

target set $\mathcal{X}_i$, $i \in \mathcal{V}$ can be also observed and considered as the self-rendezvous region. The objective is to ensure that the generalized coordinate derivatives of all the agents converge to zero and their generalized coordinates achieve agreement, while the destination of each agent is constrained by its target set.

3.4.3 Multiple Unmanned Aerial Vehicles

It is known that the formation of multiple unmanned aerial vehicles (UAVs) can be used to accomplish complex missions efficiently while the cost is largely reduced compared with a single monolithic UAV (see Fig. 3.5 for illustration). Kinematics and dynamics of a UAV can be described by

$$\dot{p}_i = v_i,$$

$$\dot{v}_i = -g\hat{e}_3 + \frac{T_i}{m_i} R_i \hat{e}_3,$$

$$\dot{R}_i = R_i S(\omega_i),$$

$$J_i \dot{\omega}_i = -S(\omega_i) J_i \omega_i + \tau_i,$$

where $p_i = [p_{i,x}, p_{i,y}, p_{i,z}]^T$ and $v_i = [v_{i,x}, v_{i,y}, v_{i,z}]^T$ denote the position and velocity of the center of gravity of UAV in inertial frame, m_i is the total mass, g is the local gravitational acceleration, $\hat{e}_3 \triangleq [0, 0, 1]^T$, T_i denotes the applied thrust along $\hat{e}_3$, R_i denotes the rotation matrix, $\omega_i = [\omega_{i,x}, \omega_{i,y}, \omega_{i,z}]^T$ denotes the angular velocity, $J_i = \text{diag}\{J_{i,x}, J_{i,y}, J_{i,z}\}$ denotes the inertial matrix, and τ_i denotes the applied torque. The objective is to guarantee that velocities of all the agents converge to the same one, and the formation converges to a well-defined and unique one with desired specified relative positions.

Fig. 3.5 Multiple UAVs

3.5 Literature

Lemma 3.3 can be easily obtained according to Theorem 4.29 in [1] and Lemma 1.6 in [2]. The dynamics of spacecraft attitudes are taken from [3], the dynamics of Lagrangian systems are taken from [4], and the dynamics of unmanned aerial vehicles are taken from [5].

References

1. Z. Qu, *Cooperative Control of Dynamical Systems: Applications to Autonomous Vehicles* (Springer, London, 2009)
2. W. Ren, Y. Cao, *Distributed Coordination of Multi-agent Networks: Emergent Problems, Models, and Issues* (Springer, London, 2011)
3. H. Schaub, J.L. Junkins, *Analytical Mechanics of Space Systems* (American Institute of Aeronautics and Astronautics, Inc, Reston, 2003)
4. M.W. Spong, S. Hutchinson, M. Vidyasagar, *Robot Dynamics and Control* (Wiley, New York, 2006)
5. Y. Zou, Z. Zhou, X. Dong et al., Distributed formation control for multiple vertical takeoff and landing UAVs with switching topologies. IEEE/ASME Trans. Mech. **23**(4), 1750–1761 (2018)

Part II
Analysis

Chapter 4
System Synchronization

This chapter considers synchronization of networked nonlinear dynamical systems under switching communication topologies. In particular, the agent dynamics are unstable and the interaction graph is time-varying with weak connectivity. We first give definitions on consensus and synchronization and present the considered problem. Then, the case of a directed graph with the uniform joint connectivity is considered. We establish a sufficient condition for reaching global exponential synchronization in terms of the relationship between the global Lipschitz constant and the network parameters. Moreover, for the case of an undirected graph with infinite joint connectivity, we show that the commonly used Lipschitz condition on the nonlinear self-dynamics is not sufficient to ensure synchronization even for an arbitrarily large coupling strength. By relaxing the global Lipschitz condition to the global Lipschitz-like condition, a sufficient synchronization condition is established in terms of the times of connectivity, the integral of the Lipschitz gain, and the network parameters.

4.1 Definitions on Consensus and Synchronization

Consensus and synchronization are commonly studied concepts in networked dynamical system (3.1). We next precisely give several relevant definitions on consensus and synchronization that will be used. Note that in (3.1), the state dimension of each agent can be different while in Chaps. 4–9 we focus on the homogenous case with the same n_i, $i \in \mathcal{V}$.

Definition 4.1 The networked dynamical system (3.1) achieves global asymptotic state consensus if for any $x_i(0) \in \mathbb{R}^n$, $i \in \mathcal{V}$, $\lim_{t \to \infty} x_i(t) = x_\star$, $\forall i \in \mathcal{V}$, where $x_\star \in \mathbb{R}^n$ is a constant vector.

© The Author(s), under exclusive license to Springer Nature Switzerland AG 2021   31
Z. Meng et al., *Modelling, Analysis, and Control of Networked Dynamical Systems*,
Systems & Control: Foundations & Applications,
https://doi.org/10.1007/978-3-030-84682-4_4

Definition 4.2 The networked dynamical system (3.1) achieves global asymptotic state synchronization if for any $x_i(0) \in \mathbb{R}^n$, $i \in \mathcal{V}$, $\lim_{t \to \infty}(x_i(t) - x_j(t)) = 0$, $\forall i, j \in \mathcal{V}$.

Definition 4.3 The networked dynamical system (3.1) achieves global exponential state synchronization if any $x_i(0) \in \mathbb{R}^n$, $i \in \mathcal{V}$, there exist constants $V_\star \geq 0$ and $\lambda > 0$ such that $\|x_i(t) - x_j(t)\| \leq e^{-\lambda t} V_\star$, $\forall i, j \in \mathcal{V}$, $\forall t \geq 0$.

Definition 4.4 The networked dynamical system (3.1) achieves global asymptotic φ-synchronization, where $\varphi : \mathbb{R}^n \to \mathbb{R}$ is a continuously differentiable function, if any $x_i(0) \in \mathbb{R}^n$, $i \in \mathcal{V}$, there exists a constant $d_\star$, such that $\lim_{t \to \infty} \varphi(x_i(t)) = d_\star$, $\forall i \in \mathcal{V}$.

Definition 4.5 The networked dynamical system (3.1) achieves global asymptotic P-norm synchronization if for any $x_i(0) \in \mathbb{R}^n$, $i \in \mathcal{V}$, there exists a constant $d_\star \geq 0$ such that

$$\lim_{t \to \infty} \|x_i(t)\|_P^2 = d_\star, \quad \forall i \in \mathcal{V}, \tag{4.1}$$

where $P = P^{\mathrm{T}} > 0$ is a symmetric positive definite matrix.

Note that consensus means that the states will eventually converge to a constant while synchronization may converge to a time-varying trajectory. On the other hand, φ-synchronization and P-norm synchronization only guarantee that the function value determined by the individual agent state reaches a consensus. In particular, Fig. 4.1 illustrates the convergence of consensus. Figure 4.2a and b illustrate the convergence of synchronization. In addition, Fig. 4.3 illustrates the convergence of φ-synchronization (and also P-norm synchronization). Considering the scalar case, we can see from Fig. 4.1 that the states of all three agents converge finally to a nonzero constant. This is consistent with the definition of consensus, i.e., Definition 4.1. According to Fig. 4.2a and b, the states of all the agents converge finally to a time-varying trajectory. This result does not satisfy Definition 4.1, but is consistent with the definition of synchronization, i.e., Definition 4.2. Finally, from Fig. 4.3, we can see that the final values of $\varphi(x_i)$, $i = 1, 2, 3$, converge to a constant, where $\varphi(x_i) = x_i^2$, for $i = 1, 2, 3$. Therefore, φ-synchronization, i.e., Definition 4.4 is guaranteed and also P-norm synchronization, i.e., Definition 4.5 is achieved with $P = 1$.

4.2   Problem Formulation

In this chapter, we consider the synchronization problem for a network of nonlinear agents with agent set $\mathcal{V}$. Their interactions are described by a time-varying directed graph $\mathcal{G}_{\sigma(t)} = (\mathcal{V}, \mathcal{E}_{\sigma(t)})$. In order to concentrate on the constraint of inherent dynamics, we specify the dynamics of (3.1) into the following equation:

Fig. 4.1 Convergence of
consensus

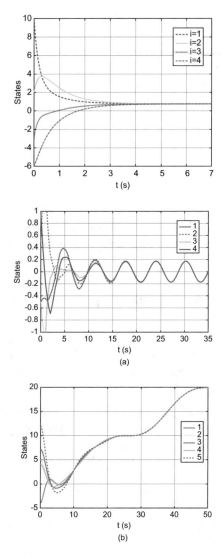

Fig. 4.2 Convergence of
synchronization. (**a**)
Convergence of
synchronization I. (**b**)
Convergence of
synchronization II

$$\dot{x}_i = f(t, x_i) + \gamma \sum_{j \in \mathcal{N}_i(\sigma(t))} a_{ij}(\sigma(t))(x_j - x_i), \quad i \in \mathcal{V}, \tag{4.2}$$

where $x_i \in \mathbb{R}^n$ is the state of agent i, $\gamma > 0$ is a coupling gain, $a_{ij}(p) > 0$ is the (i, j)th entry of the adjacency matrix A_p associated with the graph $\mathcal{G}_p$ for all $p \in \mathcal{P}$, and $f(t, x_i) : [0, \infty) \times \mathbb{R}^n \to \mathbb{R}^n$ is piecewise continuous in t and continuous in x_i representing the nonlinear self-dynamics of agent i.

First, it is not hard to show that $a_* \leq a_{ij}(p) \leq a^*$, for all $a_{ij}(p) \neq 0$, all $i, j \in \mathcal{V}$, and all $p \in \mathcal{P}$, where $a^* = \max_{p \in \mathcal{P}, i, j \in \mathcal{V}} a_{ij}(p)$ and $a_* = \min_{p \in \mathcal{P}, i, j \in \mathcal{V}} \{a | a \in$

Fig. 4.3 Convergence of
φ-synchronization and
P-norm synchronization

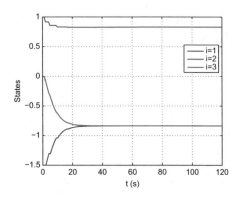

$\{a_{ij}(p)\}$ and $a \neq 0\}$. We denote $x = [x_1^T, x_2^T, \dots, x_N^T]^T \in \mathbb{R}^{Nn}$ and assume that the
initial state $x(0) = (x_1^T(0), \dots, x_N^T(0))^T \in \mathbb{R}^{Nn}$.

4.3   Directed Graph

In the existing literature, it has been established that for fixed connected graphs
[3, 10, 12] and for some special switching graphs [5, 9], synchronization is achieved
for a sufficiently large coupling γ if the self-dynamics satisfy the following global
Lipschitz assumption.

Assumption 4.1 (Also See Definition 2.3) *The self-dynamics $f(t, x_i)$, $i \in \mathcal{V}$ are
globally Lipschitz continuous in $x_i \in \mathbb{R}^n$ with the Lipschitz constant $\delta > 0$, i.e.,*

$$\| f(t, x_i) - f(t, \bar{x}_i) \| \leq \delta \| x_i - \bar{x}_i \|, \forall x_i, \bar{x}_i \in \mathbb{R}^n, \forall t \geq 0. \tag{4.3}$$

We start by considering a directed graph and focus on the case of uniform joint
strong connectivity. Before explicitly characterizing the convergence result, we first
introduce a scalar quantity and present the following lemmas in this section. Define

$$V(t, x(t)) = \max_{\{i,j\} \in \mathcal{V} \times \mathcal{V}} V_{ij}(t, x(t)), \tag{4.4}$$

where

$$V_{ij}(t, x(t)) = \frac{1}{2} e^{-2\delta t} \| x_i(t) - x_j(t) \|^2, \forall \{i, j\} \in \mathcal{V} \times \mathcal{V}. \tag{4.5}$$

Lemma 4.6 *Let Assumption 4.1 hold. Along the networked dynamical dynamics
(4.2), $V(t, x(t))$ defined in (4.4) is non-increasing for all $t \geq 0$.*

Proof This lemma establishes a critical non-expansive property along the networked dynamical dynamics (4.2) for the globally Lipschitz case. The proof follows by investigating the Dini derivative of $V(t, x(t))$.

Let $\overline{\mathcal{V}}_1 \times \overline{\mathcal{V}}_2$ be the set containing all the node pairs that reach the maximum at time t, i.e., $\overline{\mathcal{V}}_1(t) \times \overline{\mathcal{V}}_2(t) = \{\{i, j\} \in \mathcal{V} \times \mathcal{V} : V_{ij}(t) = V(t)\}$. It is not hard to obtain that

$$
\begin{aligned}
D^+ V = \max_{\{i,j\} \in \overline{\mathcal{V}}_1 \times \overline{\mathcal{V}}_2} & \Big\{ e^{-2\delta t} (x_i - x_j)^{\mathrm{T}} (f(t, x_i) - f(t, x_j)) \\
& + \gamma e^{-2\delta t} (x_i - x_j)^{\mathrm{T}} \sum_{k_1 \in \mathcal{N}_i(\sigma(t))} a_{ik_1}(\sigma(t))(x_{k_1} - x_i) \\
& - \gamma e^{-2\delta t} (x_i - x_j)^{\mathrm{T}} \sum_{k_2 \in \mathcal{N}_j(\sigma(t))} a_{jk_2}(\sigma(t))(x_{k_2} - x_j) \\
& - \delta e^{-2\delta t} \|x_i - x_j\|^2 \Big\} \\
\leq \frac{\gamma}{2} \max_{\{i,j\} \in \overline{\mathcal{V}}_1 \times \overline{\mathcal{V}}_2} & \Big\{ e^{-2\delta t} \sum_{k_1 \in \mathcal{N}_i(\sigma(t))} a_{ik_1}(\sigma(t))(\|x_j - x_{k_1}\|^2 - \|x_i - x_j\|^2) \\
& + e^{-2\delta t} \sum_{k_2 \in \mathcal{N}_j(\sigma(t))} a_{jk_2}(\sigma(t))(\|x_i - x_{k_2}\|^2 - \|x_i - x_j\|^2) \Big\} \\
\leq \gamma \max_{\{i,j\} \in \overline{\mathcal{V}}_1 \times \overline{\mathcal{V}}_2} & \Big\{ \sum_{k_1 \in \mathcal{N}_i(\sigma(t))} a_{ik_1}(\sigma(t))(V_{jk_1} - V_{ij}) \\
& + \sum_{k_2 \in \mathcal{N}_j(\sigma(t))} a_{jk_2}(\sigma(t))(V_{ik_2} - V_{ij}) \Big\} \leq 0,
\end{aligned}
\tag{4.6}
$$

where the first equality follows from Lemma 2.6 and (4.2), the first inequality follows from Assumption 4.1, i.e., (4.3) and Cauchy–Schwarz inequality, and the second inequality follows from (4.5) and the fact that $\gamma > 0$. To this end, we have shown that (4.4) is non-increasing for all $t \geq 0$.

Lemma 4.7 *Let Assumption 4.1 hold. Assume that $\mathcal{G}_{\sigma(t)}$ is uniformly jointly strongly connected (also see Definition 2.8). Then there exists $0 < \tilde{\beta} < 1$ such that*

$$
V_{ij}(\bar{N} T_0, x(\bar{N} T_0)) \leq \tilde{\beta} V_0, \qquad \forall \{i, j\} \in \mathcal{V} \times \mathcal{V},
$$

where $\bar{N} = N - 1$, $V_0 = V(0, x(0))$, and $T_0 \triangleq T + 2\tau_D$ with T given in Definition 2.8 and τ_D is the dwell time.

Proof The proof is based on the convergence analysis of $V_{ij}(t, x(t))$ for all agent pairs $\{i, j\} \in \mathcal{V} \times \mathcal{V}$ in four steps. In the following analysis, we sometimes denote $V(t, x(t))$ and $V_{ij}(t, x(t))$ as V and V_{ij}, respectively, for notational simplification.

Step 1.   We begin by considering any agent $i_1 \in \mathcal{V}$. Since $\mathcal{G}_{\sigma(t)}$ is uniformly jointly strongly connected, we know that i_1 is the root and that there exists a time t_1 and an agent $i_2 \in \mathcal{V} \setminus \{i_1\}$ such that $(i_1, i_2) \in \mathcal{E}$ during $t \in [t_1, t_1 + \tau_D) \subset [0, T_0]$. We first note that it follows from (4.4) and Lemma 4.6 that for all $t \in [0, \bar{N} T_0]$,

$$V_{ij}(t, x(t)) \leq V(t, x(t)) \leq V_0, \quad \forall \{i, j\} \in \mathcal{V} \times \mathcal{V}. \tag{4.7}$$

Taking the derivative of V_{ij} along the trajectories of (4.2), we obtain that for all $t \in [t_1, t_1 + \tau_D)$,

$$\dot{V}_{i_1 i_2} = -\delta e^{-2\delta t} \|x_{i_1} - x_{i_2}\|^2 + e^{-2\delta t} (x_{i_1} - x_{i_2})^{\mathrm{T}}$$

$$\times \Bigg\{ \gamma \sum_{k_1 \in \mathcal{N}_{i_1}(\sigma(t))} a_{i_1 k_1}(\sigma(t))(x_{k_1} - x_{i_1})$$

$$- \gamma \sum_{k_2 \in \mathcal{N}_{i_2}(\sigma(t))} a_{i_2 k_2}(\sigma(t))(x_{k_2} - x_{i_2}) + (f(t, x_{i_1}) - f(t, x_{i_2})) \Bigg\}$$

$$\leq \gamma \sum_{k_1 \in \mathcal{N}_{i_1}(\sigma(t))} a_{i_1 k_1}(\sigma(t))(V_{i_2 k_1} - V_{i_1 i_2}) - \gamma a_{i_2 i_1}(\sigma(t)) V_{i_1 i_2}$$

$$+ \gamma \sum_{k_2 \in \mathcal{N}_{i_2}(\sigma(t)) \setminus \{i_1\}} a_{i_2 k_2}(\sigma(t))(V_{i_1 k_2} - V_{i_1 i_2})$$

$$\leq (N - 1)\gamma a^*(V_0 - V_{i_1 i_2}) - \gamma a_* V_{i_1 i_2} + (N - 2)\gamma a^*(V_0 - V_{i_1 i_2})$$

$$= -\alpha \gamma \Big(V_{i_1 i_2} - \frac{(2N - 3)a^*}{\alpha} V_0\Big),$$

where $\alpha = (2N - 3)a^* + a_*$. The first inequality follows from (4.3) and (4.5), while the second inequality follows from (4.7).
It then follows that

$$V_{i_1 i_2}(t_1 + \tau_D, x(t_1 + \tau_D)) \leq \hat{\alpha}_1 V_0, \tag{4.8}$$

where $\hat{\alpha}_1 = 1 - \frac{a_*}{\alpha}(1 - e^{-\alpha \gamma \tau_D}) \in (0, 1)$.
Similarly we obtain that for all $t \in [t_1 + \tau_D, \bar{N} T_0]$, $\dot{V}_{i_1 i_2} \leq \bar{\alpha} \gamma (V_0 - V_{i_1 i_2})$, where $\bar{\alpha} = 2(N - 1)a^*$. It then follows from (4.8) that

$$V_{i_1 i_2}(t, x(t)) \leq \alpha_1^* V_0, \quad \forall t \in [t_1 + \tau_D, \bar{N} T_0], \tag{4.9}$$

where $\alpha_1^* = 1 - (1 - e^{-\alpha \gamma \tau_D}) \frac{a_*}{\alpha} e^{-\bar{\alpha} \gamma \bar{N} T_0} \in (0, 1)$.

Step 2.   Since $\mathcal{G}_{\sigma(t)}$ is uniformly jointly strongly connected, we know that there exists a time instant t_2 and an edge from $h \in \mathcal{V}_1 \triangleq \{i_1, i_2\}$ to $i_3 \in \mathcal{V} \setminus \mathcal{V}_1$ during $[t_2, t_2 + \tau_D) \subset [T_0, 2T_0]$.

We then estimate an upper bound for V_{hi_3} by considering two different cases: $h = i_1$ and $h = i_2$. We eventually obtain that for all $t \in [t_2 + \tau_D, \bar{N}T_0]$,

$$V_{hi_3}(t, x(t)) \le (1 - \beta_*^2)V_0, \quad \forall h \in \mathcal{V}_1, \tag{4.10}$$

where

$$\beta_* = (1 - e^{-\alpha\gamma\tau_D})\frac{a_*}{\alpha}e^{-\bar{\alpha}\gamma\bar{N}T_0} \in (0, 1). \tag{4.11}$$

It then follows from (4.9), (4.10), and $1 - \beta_*^2 > 1 - \beta_* = \alpha_1^*$ that for all $t \in [2T_0, \bar{N}T_0]$,

$$V_{i_1 k}(t, x(t)) \le (1 - \beta_*^2)V_0, \quad \forall k \in \mathcal{V}_2 \setminus \{i_1\},$$

where $\mathcal{V}_2 \triangleq \{i_1, i_2, i_3\}$.

Step 3.   By continuing the above process, we obtain that for all $k \in \mathcal{V} \setminus \{i_1\}$,

$$V_{i_1 k}(\bar{N}T_0, x(\bar{N}T_0)) \le (1 - \beta_*^{\bar{N}})V_0. \tag{4.12}$$

Step 4.   Since $\mathcal{G}_{\sigma(t)}$ is uniformly jointly strongly connected, (4.12) holds for any $i_1 \in \mathcal{V}$. By using the same analysis, we eventually obtain that for all $i, j \in \mathcal{V}$,

$$V_{ij}(\bar{N}T_0, x(\bar{N}T_0)) \le (1 - \beta_*^{\bar{N}})V_0.$$

Hence the result follows by choosing $\tilde{\beta} = 1 - \beta_*^{\bar{N}}$ with β_* given by (4.11). $\blacksquare$

Theorem 4.8 *Let Assumption 4.1 hold. Assume that $\mathcal{G}_{\sigma(t)}$ is uniformly jointly strongly connected. Global exponential synchronization (see Definition 4.3) is achieved for the networked dynamical system (4.2) if $\delta < \rho_*/2$, where ρ_* is a constant depending on the network parameters.*

Proof By using Lemma 4.7 and (4.4), we obtain that

$$V(t, x(t)) \le \tilde{\beta}^{\lfloor \frac{t}{\bar{N}T_0} \rfloor}V_0 \le \frac{1}{\tilde{\beta}}e^{-\rho_* t}V_0,$$

where $\lfloor \frac{t}{\bar{N}T_0} \rfloor$ denotes the largest integer that is not greater than $\frac{t}{\bar{N}T_0}$ and $\rho_* = \frac{1}{\bar{N}T_0}\ln\frac{1}{\tilde{\beta}}$.

It then follows from (4.4) and (4.5) that

$$\max_{\{i,j\}\in\mathcal{V}\times\mathcal{V}} \|x_i(t) - x_j(t)\|^2 \le \frac{1}{\tilde{\beta}} e^{-(\rho_* - 2\delta)t} \max_{\{i,j\}\in\mathcal{V}\times\mathcal{V}} \|x_i(0) - x_j(0)\|^2.$$

Hence, based on the Definition 4.3, global exponential synchronization is achieved with $\xi = \frac{1}{\tilde{\beta}}$ and $\lambda = \rho_* - 2\delta$ provided that $\rho_* > 2\delta$.

4.4   Undirected Graph

For single integrators (i.e., $f(t, x_i) = 0$ in (4.2)) over an undirected graph, we know that the following assumption on infinite joint connectivity for an undirected graph (also see Definition 2.11) is a sufficient and necessary condition for achieving global asymptotic synchronization [4].

We first show that infinite joint connectivity is also necessary for achieving global asymptotic synchronization of the general networked dynamical system (4.2).

Theorem 4.9 *Assume that the equilibrium point $x = x^*$ of $\dot{x} = f(t, x)$ is not asymptotically stable. Also assume that $\mathcal{G}_{\sigma(t)}$ is undirected for all $t \ge 0$. If global asymptotic synchronization is achieved for the networked dynamical system (4.2), then $\mathcal{G}_{\sigma(t)}$ is infinitely jointly connected.*

Proof We prove Theorem 4.9 by contraposition. Suppose that $\mathcal{G}_{\sigma(t)}$ is not infinitely jointly connected. Then there exists $t^* \ge 0$ such that the union graph $\mathcal{G}([t^*, \infty))$ is not connected. This implies that there exist two nonempty, disjoint subsets $\mathcal{V}_a \subset \mathcal{V}$ and $\mathcal{V}_b \subset \mathcal{V}$ such that there is no link between sets $\mathcal{V}_a$ and $\mathcal{V}_b$ for all $t \ge t^*$. Let us choose $x_i(t^*) = x^*$ for all $i \in \mathcal{V}_a$ and $x_j(t^*) \ne x^*$ for all $j \in \mathcal{V}_b$, where $x = x^*$ is the equilibrium point of $\dot{x} = f(t, x)$. Then $x_i(t) \equiv x^*$ for all $i \in \mathcal{V}_a$ and for all $t \ge t^*$. In addition, it holds that $\dot{x}_j(t) = f(t, x_j(t))$, for all $j \in \mathcal{V}_b$ and for all $t \ge t^*$. Based on the fact that the equilibrium point $x = x^*$ of $\dot{x} = f(t, x)$ is not asymptotically stable, we know that $\lim_{t\to\infty}(x_j(t) - x^*) \ne 0$ for all $j \in \mathcal{V}_b$. This implies that global asymptotic synchronization cannot be achieved for (4.2). Hence, the result follows.

It is then natural to ask whether Assumption 4.1 is sufficient for achieving global asymptotic synchronization of (4.2) provided that $\mathcal{G}_{\sigma(t)}$ is infinitely jointly connected. We present the following example to indicate that such a consideration is not correct.

Example 4.10 Consider a group of two agents switching between two graphs $\mathcal{G}_1$ and $\mathcal{G}_2$ with adjacency matrices $A_1 = \begin{bmatrix} 0 & 0 \\ 0 & 0 \end{bmatrix}$ and $A_2 = \begin{bmatrix} 0 & 1 \\ 1 & 0 \end{bmatrix}$, respectively. The self-dynamics are $f(t, x_i) = \delta x_i$, where x_i is a scalar, $i = 1, 2$. The dynamics of each system are described by

$$\dot{x}_1 = \delta x_1 + \gamma a_{12}(\sigma(t))(x_2 - x_1),$$

$$\dot{x}_2 = \delta x_2 + \gamma a_{21}(\sigma(t))(x_1 - x_2).$$

Note that $a_{12}(\sigma(t)) = a_{21}(\sigma(t)) = 1$ or 0 for all $t \geq 0$. Then, the relative dynamics can be written as

$$\dot{\bar{x}} = (\delta - 2\gamma a_{12}(\sigma(t)))\bar{x}, \tag{4.14}$$

where $\bar{x} = x_1 - x_2$. Let the switching signal $\sigma(t)$ be equal to 2 when $t \in [\varrho^2 - 1, \varrho^2)$ and equal to 1 when $t \in [\varrho^2, (\varrho + 1)^2 - 1)$ for $\varrho = 1, 2, \ldots$. It is easy to see that the assumption on infinite joint connectivity is satisfied. It follows that the solution of (4.14) is

$$\bar{x}(t) = e^{(\delta - 2\gamma)(\rho - 1) + \delta(\rho + 1)\rho} e^{(\delta - 2\gamma)(t - \varrho^2 + 1)}\bar{x}(0)$$

for $t \in [\varrho^2 - 1, \varrho^2)$, and

$$\bar{x}(t) = e^{(\delta - 2\gamma)\rho + \delta(\rho - 1)\rho} e^{(\delta - 2\gamma)(t - \varrho^2)}\bar{x}(0)$$

for $t \in [\varrho^2, (\varrho + 1)^2 - 1)$. Thus $\lim_{t \to \infty} \bar{x}(t) = \infty$ for any $\delta > 0$ and arbitrarily large γ. Hence, even if Assumption 4.1 is satisfied, global asymptotic synchronization still cannot be achieved with arbitrarily large γ.

Motivated by the above observations, we relax Assumption 4.1 of agent dynamics to guarantee global asymptotic synchronization of (4.2) when $\mathcal{G}_{\sigma(t)}$ is infinitely jointly connected. In particular, the following global Lipschitz-like assumption is proposed.

Assumption 4.2 *There exists a continuous non-negative bounded function $\delta(t) \geq 0$ such that $\|f(t, x_i) - f(t, \bar{x}_i)\| \leq \delta(t)\|x_i - \bar{x}_i\|$, $\forall x_i, \bar{x}_i \in \mathbb{R}^n$, $\forall t \geq 0$.*

We also introduce the concept of times of connectivity. To do so, for the case of a switching graph $\mathcal{G}_{\sigma(t)}$, we first introduce a subsequence of the switching time sequence $\{t_\ell\}_0^\infty$ as $0 = T_0 < T_1 < T_2 \ldots$, where $T_\ell, \ell = 1, 2, \ldots$ is iteratively obtained by

$$T_\ell = \inf\{t \geq T_{\ell-1} : \mathcal{G}([T_{\ell-1}, t)) \text{ is connected, } T_{\ell-1} \in \{t_\ell\}_0^\infty\}.$$

Let $J(t)$ denote the number of jointly connected graphs during $[0, t)$, i.e., $J(t) = \max\{\ell : t > T_\ell\}$.

We are now ready to present a sufficient synchronization condition in terms of the times of connectivity $J(t)$ and the integral of the Lipschitz gain $\delta(t)$.

Theorem 4.11 *Let Assumption 4.2 hold. Assume that $\mathcal{G}_{\sigma(t)}$ is undirected for all $t \geq 0$ and infinitely jointly connected. Global asymptotic synchronization is achieved for the networked dynamical system (4.2) if*

$$\lim_{t\to\infty}\left(J(t)-\frac{2}{\rho}\int_0^t \delta(s)ds\right)=\infty, \tag{4.15}$$

where ρ is a constant depending on the network parameters.

Proof The proof is based on the convergence analysis of the scalar quantity

$$V(t,x)=\max_{\{i,j\}\in\mathcal{V}\times\mathcal{V}}V_{ij}(t,x), \tag{4.16}$$

where

$$V_{ij}(t,x)=\frac{1}{2\gamma}e^{-2\int_0^t \delta(s)ds}\|x_i(t)-x_j(t)\|^2. \tag{4.17}$$

We first show that along solutions to (4.2), $D^+V(t,x)\le 0$ for all $t\ge 0$.

In particular, let $\overline{\mathcal{V}}_1(t)\times\overline{\mathcal{V}}_2(t)$ be the set containing all the node pairs that reach the maximum at time t, i.e., $\overline{\mathcal{V}}_1(t)\times\overline{\mathcal{V}}_2(t)=\{\{i,j\}\in\mathcal{V}\times\mathcal{V}|V_{ij}(t)=V(t)\}$. It is not hard to obtain that

$$
\begin{aligned}
D^+V=\max_{\{i,j\}\in\overline{\mathcal{V}}_1\times\overline{\mathcal{V}}_2}&\left\{\frac{1}{\gamma}e^{-2\int_0^t \delta(s)ds}(x_i-x_j)^{\mathrm{T}}(f(t,x_i)-f(t,x_j))\right.\\
&-e^{-2\int_0^t \delta(s)ds}(x_i-x_j)^{\mathrm{T}}\sum_{k_1\in\mathcal{N}_i(\sigma(t))}a_{ik_1}(\sigma(t))(x_i-x_{k_1})\\
&+e^{-2\int_0^t \delta(s)ds}(x_i-x_j)^{\mathrm{T}}\sum_{k_2\in\mathcal{N}_j(\sigma(t))}a_{jk_2}(\sigma(t))(x_j-x_{k_2})\\
&\left.-\frac{1}{\gamma}\delta(t)e^{-2\int_0^t \delta(s)ds}\|x_i-x_j\|^2\right\}\\
\le-\frac{1}{2}e^{-2\int_0^t \delta(s)ds}&\max_{\{i,j\}\in\overline{\mathcal{V}}_1\times\overline{\mathcal{V}}_2}\left\{\sum_{k_1\in\mathcal{N}_i(\sigma(t))}a_{ik_1}(\sigma(t))\right.\\
\times(\|x_i-x_j\|^2-&\|x_j-x_{k_1}\|^2)\\
&\left.+\sum_{k_2\in\mathcal{N}_j(\sigma(t))}a_{jk_2}(\sigma(t))(\|x_j-x_i\|^2-\|x_i-x_{k_2}\|^2)\right\}\\
\le-\gamma\max_{\{i,j\}\in\overline{\mathcal{V}}_1\times\overline{\mathcal{V}}_2}&\left\{\sum_{k_1\in\mathcal{N}_i(\sigma(t))}a_{ik_1}(\sigma(t))(V_{ij}-V_{jk_1})\right.
\end{aligned}
$$

$$+ \sum_{k_2 \in \mathcal{N}_j(\sigma(t))} a_{jk_2}(\sigma(t))(V_{ij} - V_{ik_2}) \Bigg\} \leq 0,$$

where the equality follows from Lemma 2.6 and (4.2), the first inequality follows from Assumption 4.2 and Cauchy–Schwarz inequality, and the last inequality follows from (4.17) and the fact that $\gamma > 0$. Therefore, $V(t, x(t)) \leq V(0, x(0)) \triangleq V_0$.

For any node $i_1 \in \mathcal{V}$, let us define a constant $\bar{t}_1 \geq 0$ as

$$\bar{t}_1 = \inf\{t \geq 0 : \exists i_2, \text{ such that } \{i_1, i_2\} \in \mathcal{E}_{\sigma(t)}\}.$$

Note that $\bar{t}_1 + \tau_D \leq T_1$, where T_1 is defined after Assumption 4.2. Then, for $t \in [\bar{t}_1, \bar{t}_1 + \tau_D)$, it follows that

$$\dot{V}_{i_1 i_2} \leq -\gamma \sum_{k_1 \in \mathcal{N}_{i_1}(\sigma(t)) \setminus \{i_2\}} a_{i_1 k_1}(\sigma(t))(V_{i_1 i_2} - V_{i_2 k_1})$$

$$-\gamma \sum_{k_2 \in \mathcal{N}_{i_2}(t) \setminus \{i_1\}} a_{i_2 k_2}(\sigma(t))(V_{i_1 i_2} - V_{i_1 k_2})$$

$$- a_{i_1 i_2}(\sigma(t))\gamma V_{i_1 i_2} - a_{i_2 i_1}(\sigma(t))\gamma V_{i_1 i_2}$$

$$\leq -\alpha\gamma(V_{i_1 i_2}(t) - \frac{2(N-2)a^*}{\alpha} V_0),$$

where a_* and a^* are given in Sect. 4.2 and $\alpha = 2(N-2)a^* + 2a_*$. Therefore, we obtain

$$V_{i_1 i_2}(\bar{t}_1 + \tau_D) \leq \beta_1 V_0, \qquad (4.18)$$

where

$$\beta_1 = 1 - \frac{2a_*}{\alpha}(1 - e^{-\alpha\gamma\tau_D}) \in (0, 1). \qquad (4.19)$$

We next define that

$$\bar{t}_2 = \inf\{t \geq \bar{t}_1 : \exists i_3, \text{ s.t. } \{i_1, i_3\} \in \mathcal{E}_{\sigma(t)} \text{ or } \{i_2, i_3\} \in \mathcal{E}_{\sigma(t)}\}.$$

It follows from this definition that there is no edge between the set $\{i_1, i_2\}$ and the set $\mathcal{V} \setminus \{i_1, i_2\}$ for $t \in [\bar{t}_1 + \tau_D, \bar{t}_2]$. It is then not hard to see that $\dot{V}_{i_1 i_2}(t) \leq 0$ for all $t \in [\bar{t}_1 + \tau_D, \bar{t}_2]$. This together with (4.18) implies that

$$V_{i_1 i_2}(\bar{t}_2) \leq \beta_1 V_0. \qquad (4.20)$$

Note that for all $t \in [\bar{t}_2, \bar{t}_2 + \tau_D]$,

$$\dot{V}_{i_1 i_2} \leq -2(N-1)a^* \gamma (V_{i_1 i_2} - V_0). \tag{4.21}$$

By using the above relation, (4.19) and (4.20), we obtain that for all $t \in [\bar{t}_2, \bar{t}_2 + \tau_D]$,

$$V_{i_1 i_2}(t) \leq \bar{\beta} V_0, \tag{4.22}$$

where

$$\bar{\beta} = 1 - (1 - e^{-\alpha \gamma \tau_D}) \frac{2a_*}{\alpha} e^{-\alpha_1 \gamma \tau_D}, \tag{4.23}$$

and $\alpha_1 = 2(N-1)a^*$.

We next estimate $V_{i_1 i_3}$ by considering two cases.

- Case I: $(i_1, i_3) \in \mathcal{E}_{\sigma(\bar{t}_2)}$. Following a similar analysis to obtain (4.18) for $V_{i_1 i_2}$, we obtain

$$V_{i_1 i_3}(\bar{t}_2 + \tau_D) \leq \beta_1 V_0. \tag{4.24}$$

- Case II: $(i_1, i_3) \notin \mathcal{E}_{\sigma(\bar{t}_2)}$. By the definition of $\bar{t}_2$, we know that $(i_2, i_3) \notin \mathcal{E}_{\sigma(\bar{t}_2)}$. It then follows that for all $t \in [\bar{t}_2, \bar{t}_2 + \tau_D)$,

$$\dot{V}_{i_1 i_3}(t) \leq -2(N-1)a^* \gamma (V_{i_1 i_3} - V_0) - a_{i_3 i_2} \gamma (V_{i_1 i_3} - V_{i_1 i_2}). \tag{4.25}$$

We proceed our analysis for two subcases.

- Case II(a): $V_{i_1 i_3}(t) > V_{i_1 i_2}(t)$ for all $t \in [\bar{t}_2, \bar{t}_2 + \tau_D)$. It then follows that

$$\dot{V}_{i_1 i_3}(t) \leq -\alpha_2 \gamma (V_{i_1 i_3} - \frac{2(N-1)a^* + a_* \bar{\beta}}{\alpha_2} V_0),$$

where $\alpha_2 = 2(N-1)a^* + a_*$. This shows that

$$V_{i_1 i_3}(\bar{t}_2 + \tau_D) \leq \left(1 - \frac{a_*(1-\bar{\beta})}{\alpha_2}(1 - e^{-\alpha_2 \gamma \tau_D})\right) V_0. \tag{4.26}$$

- Case II(b): there exists a time $t^* \in [\bar{t}_2, \bar{t}_2 + \tau_D)$ such that

$$V_{i_1 i_3}(t^*) \leq V_{i_1 i_2}(t^*) \leq \bar{\beta} V_0. \tag{4.27}$$

Applying the same analysis as we obtained (4.21) to (4.25) yields,

$$\dot{V}_{i_1 i_3}(t) \leq -2(N-1)a^* \gamma (V_{i_1 i_3} - V_0). \tag{4.28}$$

By using (4.27), (4.28), and $\alpha_1 < \alpha_2$, we obtain that for all $t \in [t^*, \bar{t}_2 + \tau_D)$,

$$V_{i_1 i_3}(\bar{t}_2 + \tau_D) \le \left(1 - e^{-\alpha_2 \gamma \tau_D}(1 - \bar{\beta})\right) V_0. \tag{4.29}$$

We shall find an upper bound for $V_{i_1 i_3}$ for the above cases. It follows from (4.23) and (4.29) that for Case II(b),

$$V_{i_1 i_3}(\bar{t}_2 + \tau_D) \le \left(1 - e^{-(\alpha_1 + \alpha_2)\gamma \tau_D}(1 - e^{-\alpha \gamma \tau_D})\frac{2a_*}{\alpha}\right) V_0.$$

Also note that from (4.19) and (4.24), for Case I, we have

$$V_{i_1 i_3}(\bar{t}_2 + \tau_D) \le \left(1 - (1 - e^{-\alpha \gamma \tau_D})\frac{2a_*}{\alpha}\right) V_0.$$

Therefore, the bound in Case II(b) is larger than that of Case I. By noting that $\frac{a_*(1-\bar{\beta})}{\alpha_2}(1 - e^{-\alpha_2 \gamma \tau_D})e^{-\alpha_2 \gamma \tau_D} \le \min\{e^{-\alpha_2 \gamma \tau_D}(1 - \bar{\beta}), \frac{a_*(1-\bar{\beta})}{\alpha_2}(1 - e^{-\alpha_2 \gamma \tau_D})\}$ and comparing (4.26) and (4.29), it is not hard to see that

$$V_{i_1 i_3}(\bar{t}_2 + \tau_D) \le \left(1 - \frac{a_*(1 - \bar{\beta})}{\alpha_2}(1 - e^{-\alpha_2 \gamma \tau_D})e^{-\alpha_2 \gamma \tau_D}\right) V_0 = \beta_2 V_0, \tag{4.30}$$

where

$$\beta_2 = 1 - \frac{2a_*^2}{\alpha \alpha_2}(1 - e^{-\alpha_2 \gamma \tau_D})e^{-\alpha_2 \gamma \tau_D}(1 - e^{-\alpha \gamma \tau_D})e^{-\alpha_1 \gamma \tau_D} \in (0, 1).$$

Note that $\bar{t}_2 + \tau_D \le T_2$. It follows from (4.22) and (4.30) that

$$V_{i_1 k}(\bar{t}_2 + \tau_D) \le \beta_2 V_0, \ k \in \{i_2, i_3\}.$$

We then proceed the above analysis for other nodes $k \in \mathcal{V} \backslash \{i_1\}$. Eventually, we obtain that

$$V_{i_1 k}(\bar{t}_{N-1} + \tau_D) \le \beta_{N-1} V_0, \ k \in \mathcal{V} \backslash \{i_1\},$$

where $\bar{t}_{N-1} + \tau_D \le T_{N-1}$.

Let us now consider node i_2 and try to bound $V_{i_2 i_3}, \ldots, V_{i_2 i_N}$. By going through a similar analysis, we obtain that $V_{i_2 k}(\bar{t}_N + \tau_D) \le \beta_{2N-3} V_0, \ k \in \mathcal{V} \backslash \{i_2\}$. We can eventually obtain that

$$V_{ij}(\bar{t}_{(N-1)N/2} + \tau_D) \le \beta_{(N-1)N/2} V_0, \ \forall i, j \in \mathcal{V},$$

where $\beta_{(N-1)N/2}$ is a constant depending on the network parameters, namely, τ_D, N, a^*, a_*, and γ. Also note that $\bar{t}_{(N-1)N/2} + \tau_D \le T_{(N-1)N/2}$. Therefore, it follows that $V(T_{(N-1)N/2+1}) \le \beta_{(N-1)N/2} V_0$. We then have that

$$V(t, x(t)) \le \beta_{(N-1)N/2}^{\lfloor \frac{J(t)}{(N-1)N/2+1} \rfloor} V_0 \le \frac{1}{\beta_{(N-1)N/2}} e^{-\rho J(t)} V_0, \tag{4.31}$$

and

$$\rho = \frac{1}{(N-1)N/2+1} \ln \frac{1}{\beta_{(N-1)N/2}} > 0. \tag{4.32}$$

It thus follows from (4.16), (4.17), and (4.31) that

$$\max_{\{i,j\} \in \mathcal{V} \times \mathcal{V}} \|x_i(t) - x_j(t)\|^2 \le \frac{2\gamma}{\beta_{(N-1)N/2}} e^{2 \int_0^t \delta(s)\mathrm{d}s - \rho J(t)} V_0.$$

Hence, global asymptotic synchronization is achieved provided that

$$\lim_{t \to \infty} \left(J(t) - \frac{2}{\rho} \int_0^t \delta(s)\mathrm{d}s \right) = \infty.$$

Motivated by the above theorem, we can slightly revise Example 4.10 so that global asymptotic synchronization is achieved.

Example 4.12 Let the switching signal $\sigma(t)$ be equal to 2 when $t \in [\varrho^2 - 1, \varrho^2)$ and equal to 1 when $t \in [\varrho^2, (\varrho+1)^2 - 1)$ for $\varrho = 1, 2, \ldots$, but the self-dynamics now be $f(t, x_i) = \frac{\delta x_i}{t}, i = 1, 2$, where x_i is scalar. Then, global asymptotic synchronization is achieved if $\delta < 1$.

4.5 Simulations

In this section, we present two examples on φ-synchronization to illustrate our results.

4.5.1 Directed Graph

In the first example, we consider the multi-agent system (4.2) with $x_i \in \mathbb{R}^3$ and $f(t, x_i) = A x_i, i = 1, 2, 3, 4$. In particular, $A = \begin{bmatrix} 0.05 & 0.05 & 0.05 \\ 0.05 & 0.05 & 0.05 \\ 1 & 1 & 0 \end{bmatrix}$, and $\gamma = 5$. It is not hard to see that Assumption 4.3 is satisfied.

The switching topology $\mathcal{G}_{\sigma(t)}$ is defined by the piecewise constant switching signal

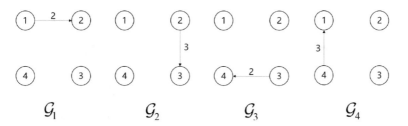

Fig. 4.4 Switching topology $\mathcal{G}_{\sigma(t)}$

Fig. 4.5 State convergence
for Theorem 4.8

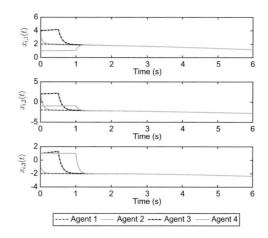

$$\sigma(t) = \begin{cases} 1, & t \in [2\ell, 2\ell + 0.5), \\ 2, & t \in [2\ell + 0.5, 2\ell + 1), \\ 3, & t \in [2\ell + 1, 2\ell + 1.5), \\ 4, & t \in [2\ell + 1.5, 2\ell + 2), \end{cases}$$

where $\ell = 0, 1, \ldots$. The directed switching graph is defined by $\mathcal{G}_i$ for $i = 1, \ldots, 4$ in Fig. 4.4. Note that $\mathcal{G}([2\ell, 2\ell + 2))$ is uniformly jointly strongly connected, and hence satisfies the condition in Theorem 4.8. Figure 4.5 shows that global asymptotic synchronization is achieved, which agrees with the results of Theorem 4.8.

4.5.2 Undirected Graph

In the second example, we still consider the multi-agent system (4.2) with $x_i \in \mathbb{R}$ and $f(t, x_i) = \frac{0.1 x_i}{t}$, $i = 1, 2, 3, 4$. In addition, $\gamma = 2$. It is not hard to see that Assumption 4.2 is satisfied. The switching topology $\mathcal{G}_{\sigma(t)}$ is defined by the piecewise constant switching signal

Fig. 4.6 Switching topology $\mathcal{G}_{\sigma(t)}$

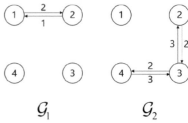

Fig. 4.7 State convergence for Theorem 4.11

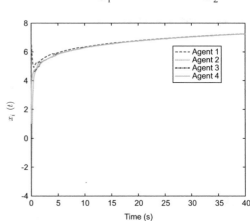

$$\sigma(t) = \begin{cases} 1, & t \in [\ell^2, \ell^2 + 0.5), \\ 2, & \text{otherwise,} \end{cases}$$

where $\ell = 0, 1, \ldots$. The switching graph is given by $\mathcal{G}_1$ and $\mathcal{G}_2$ in Fig. 4.6. Note that $\mathcal{G}_{\sigma(t)}$ is infinitely jointly connected, and hence satisfies the condition in Theorem 4.11. Figure 4.7 shows that global asymptotic synchronization is achieved, which agrees with the results of Theorem 4.11.

4.6 Literature

When the communication constraint is considered, the current literature mainly focuses on the case where the communication graph is fixed. It is shown that for the case where the agent nonlinear dynamics satisfy a Lipschitz condition, synchronization is achieved for a connected graph provided that the coupling strength is sufficiently large. The results in this chapter are based mainly on [11]. This chapter studies synchronization of general networked nonlinear dynamical systems under time-varying communication topologies.

For the case where the communication graph is time-varying, the synchronization problem becomes much more challenging and existing literature mainly focuses on a few special cases when the nonlinear agent dynamics become linear, e.g., the single-integrator case [1, 2, 4], the double-integrator case [6], and the neutrally stable case [7, 8]. Other studies assume some particular structures for the communication graph [5, 9, 14]. In particular, in [13], the authors focus on the case where the adjacency matrices associated with all communication graphs are simultaneously triangularizable. The authors of [5] consider switching communication graphs that are weakly connected and balanced at all times. A more general case where the switching communication graph frequently has a directed spanning tree has been considered in [9]. These special structures on the switching communication graph are restrictive compared to joint connectivity where the communication can be lost at any time.

Acknowledgments ©2016 IEEE. Reprinted, with permission, from Tao Yang, Ziyang Meng, Wei Ren, Karl H. Johansson, "Synchronization of coupled nonlinear dynamical systems: interplay between times of connectivity and integral of Lipschitz gain", IEEE Transactions on Circuits and Systems II: Berif Papers, vol. 63, no. 4, pp. 391–395, 2016.

References

1. A. Jadbabaie, J. Lin, A.S. Morse, Coordination of groups of mobile autonomous agents using nearest neighbor rules. IEEE Trans. Autom. Control **48**(6), 988–1001 (2003)
2. Z. Lin, B. Francis, M. Maggiore, State agreement for continuous-time coupled nonlinear systems. SIAM J. Control Optim. **46**(1), 288–307 (2007)
3. H. Liu, M. Cao, C.W. Wu, Coupling strength allocation for synchronization in complex networks using spectral graph theory. IEEE Trans. Circ. Syst. I Regul. Pap. **61**(5), 1520–1530 (2014)
4. L. Moreau, Stability of multi-agent systems with time-dependent communication links. IEEE Trans. Autom. Control **50**(2), 169–182 (2005)
5. J. Qin, H. Gao, W.X. Zheng, Exponential synchronization of complex networks of linear systems and nonlinear oscillators: a unified analysis. IEEE Trans. Neural Netw. Learn. Syst. **26**(3), 510–521 (2014)
6. W. Ren, R.W. Beard, Consensus seeking in multiagent systems under dynamically changing interaction topologies. IEEE Trans. Autom. Control **50**(5), 655–661 (2005)
7. L. Scardovi, R. Sepulchre, Synchronization in networks of identical linear systems. Automatica **45**, 546–551 (2009)
8. Y. Su, J. Huang, Cooperative output regulation of linear multi-agent systems. IEEE Trans. Autom. Control **57**(4), 1062–1066 (2012)
9. G. Wen, Z. Duan, G. Chen et al. Consensus tracking of multi-agent systems with Lipschitz-type node dynamics and switching topologies. IEEE Trans. Circ. Syst. I Regul. Pap. **61**(2), 499–511 (2014)
10. C.W. Wu, *Synchronization in Complex Networks of Nonlinear Dynamical Systems* (World Scientific, Singapore, 2007)
11. T. Yang, Z. Meng, W. Ren et al., Synchronization of coupled nonlinear dynamical systems: interplay between times of connectivity and integral of Lipschitz gain. IEEE Trans. Circ. Syst. II Express Briefs **63**(4), 391–395 (2015)
12. W. Yu, G. Chen, M. Cao, Consensus in directed networks of agents with nonlinear dynamics. IEEE Trans. Autom. Control **56**(6), 1436–1441 (2011)

13. J. Zhao, D.J. Hill, T. Liu, Passivity-based output synchronization of dynamical network with non-identical nodes, in *Proceedings of the IEEE Conference on Decision and Control*, Atlanta, 2010, pp. 7351–7356
14. J. Zhao, D.J. Hill, T. Liu, Synchronization of dynamical networks with nonidentical nodes: criteria and control. IEEE Trans. Circ. Syst. I Regul. Pap. **58**(3), 584–594 (2011)

Chapter 5
Synchronization with Non-expansive Dynamics

In Chap. 4, we have studied state synchronization for networked nonlinear dynamical systems under switching communication topologies. Sufficient conditions are established in terms of connectivity, nonlinear self-dynamics, and the network parameters. On the other hand, a stronger condition can be obtained for the case where the nonlinear agent dynamics are non-expansive dynamics, i.e., stable dynamics with a convex Lyapunov function $\varphi(\cdot)$. In particular, we show that several forms of joint connectivity for communication graphs are sufficient for networks to achieve global asymptotic φ-synchronization in this chapter.

5.1 Problem Formulation

Consider a network with N networked nonlinear dynamical systems as studied in Chap. 4. The dynamics of the systems are described by the model (4.2). We know from Sect. 4.2 that $a_* \leq a_{ij}(p) \leq a^*$, for all $a_{ij}(p) \neq 0$, all $i, j \in \mathcal{V}$, and all $p \in \mathcal{P}$, where a^* and a_* are defined in Sect. 4.2.

We denote in this chapter $x = [x_1^\mathrm{T}, x_2^\mathrm{T}, \ldots, x_N^\mathrm{T}]^\mathrm{T} \in \mathbb{R}^{Nn}$ and the initial state $x(0) = (x_1^\mathrm{T}(0), \ldots, x_N^\mathrm{T}(0))^\mathrm{T} \in \mathbb{R}^{Nn}$. Also, we are interested in global asymptotic φ-synchronization as defined in Definition 4.4.

In this chapter, we focus on the case that the nonlinear inherent agent dynamics are non-expansive as indicated by the following assumption.

Assumption 5.1 *There exists $\varphi : \mathbb{R}^n \to \mathbb{R}$ that is a continuously differentiable positive definite convex function satisfying*

(i) $\lim_{\|\eta\| \to \infty} \varphi(\eta) = \infty$;

(ii) $\langle \nabla\varphi(\eta), f(t, \eta) \rangle \leq 0$ *for any $\eta \in \mathbb{R}^n$ and any $t \geq 0$.*

© The Author(s), under exclusive license to Springer Nature Switzerland AG 2021
Z. Meng et al., *Modelling, Analysis, and Control of Networked Dynamical Systems*,
Systems & Control: Foundations & Applications,
https://doi.org/10.1007/978-3-030-84682-4_5

5.2 Directed Graph

In this section, the directed graph case is considered and the following lemma is first presented to show how Assumption 5.1 enforces non-expansive dynamics.

Lemma 5.1 *Let Assumption 5.1 hold. Along the networked dynamical system (4.2), $\max_{i \in \mathcal{V}} \varphi(x_i(t))$ is non-increasing for all $t \geq 0$.*

Proof Denote $\overline{\mathcal{V}}(t) = \{i \in \mathcal{V} : \varphi(x_i(t)) = \max_{i \in \mathcal{V}} \varphi(x_i(t))\}$. We first note that the convexity property of $\varphi(\cdot)$ implies that [2, pp. 69]

$$\langle \nabla \varphi(\eta), \zeta - \eta \rangle \leq \varphi(\zeta) - \varphi(\eta), \quad \forall \eta, \zeta \in \mathbb{R}^n. \tag{5.1}$$

It then follows from Lemma 2.6, Assumption 5.1(ii) and (5.1) that

$$D^+ \max_{i \in \mathcal{V}} \varphi(x_i(t)) = \max_{i \in \overline{\mathcal{V}}(t)} \left\langle \nabla \varphi(x_i), f(t, x_i) + \gamma \sum_{j \in \mathcal{N}_i(\sigma(t))} a_{ij}(\sigma(t))(x_j - x_i) \right\rangle$$

$$\leq \gamma \max_{i \in \overline{\mathcal{V}}(t)} \sum_{j \in \mathcal{N}_i(\sigma(t))} a_{ij}(\sigma(t))(\varphi(x_j) - \varphi(x_i)) \leq 0,$$

where the last inequality follows from $\varphi(x_j) \leq \varphi(x_i)$ and $\gamma > 0$.

We next present the main results of the directed graph with uniform joint strong connectivity.

Theorem 5.2 *Let Assumption 5.1 hold. The networked dynamical system (4.2) achieves global asymptotic φ-synchronization if $\mathcal{G}_{\sigma(t)}$ is uniformly jointly strongly connected.*

Proof It follows from Lemma 5.1 that for any initial state $x(0) \in \mathbb{R}^{nN}$, there exists a constant $d_\star = d_\star(x(0)) \geq 0$, such that $\lim_{t \to \infty} \max_{i \in \mathcal{V}} \varphi(x_i(t)) = d_\star$. We next show that $d_\star$ is exactly the constant in Definition 4.4 of φ-synchronization.

First, it follows from Lemma 5.1 that for all $i \in \mathcal{V}$, there exist constants $0 \leq \alpha_i \leq \beta_i \leq d_\star$, such that

$$\liminf_{t \to \infty} \varphi(x_i(t)) = \alpha_i, \quad \limsup_{t \to \infty} \varphi(x_i(t)) = \beta_i.$$

Also note that it follows from the fact $\lim_{t \to \infty} \max_{i \in \mathcal{V}} \varphi(x_i(t)) = d_\star$ that for any $\varepsilon > 0$, there exists $T_1(\varepsilon) > 0$ such that

$$\varphi(x_i(t)) \in [0, d_\star + \varepsilon], \quad \forall i \in \mathcal{V}, \forall t \geq T_1(\varepsilon). \tag{5.2}$$

The proof of Theorem 5.2 is based on a contradiction argument and relies on the following lemma.

Lemma 5.3 *Let Assumption 5.1 hold. Assume that $\mathcal{G}_{\sigma(t)}$ is uniformly jointly strongly connected. If there exists an agent $k_0 \in \mathcal{V}$ such that $0 \le \alpha_{k_0} < d_\star$, then there exist $0 < \bar{\rho} < 1$ and $\bar{t}$ such that for all $i \in \mathcal{V}$, $\varphi(x_i(\bar{t} + (N-1)T_0)) \le \bar{\rho}M_0 + (1-\bar{\rho})(d_\star + \varepsilon)$, where*

$$T_0 \triangleq T + 2\tau_D, \tag{5.3}$$

with T given in Definition 2.8 and τ_D is the dwell time.

Proof Let us first define $M_0 \triangleq \frac{\alpha_{k_0} + \beta_{k_0}}{2} < d_\star$. Then there exists an infinite time sequence $0 < \tilde{t}_1 < \ldots < \tilde{t}_k < \ldots$ with $\lim_{k \to \infty} \tilde{t}_k = \infty$ such that $\varphi(x(\tilde{t}_k)) = M_0$ for all $k = 1, 2, \ldots$. We then pick up one $\tilde{t}_k$, $k = 1, 2, \ldots$ such that it is greater than or equal to $T_1(\varepsilon)$ and denote it as $\tilde{t}_{k_0}$.

We now prove this lemma by estimating an upper bound of the scalar function $\varphi(x_i)$ agent by agent. The proof is based on a generalization of the method proposed in the proof of [18, Lemma 4.3] but with substantial differences on the agent dynamics and Lyapunov function. Moreover, the convexity of $\varphi(\cdot)$ plays an important role.

Step 1.   We focus on agent k_0. By using Assumption 5.1(ii), (5.1), and (5.2), we obtain that for all $t \ge \tilde{t}_{k_0}$,

$$\frac{d}{dt}\varphi(x_{k_0}(t)) = \left\langle \nabla\varphi(x_{k_0}), f(t, x_{k_0}) + \gamma \sum_{j \in \mathcal{N}_{k_0}(\sigma(t))} a_{k_0 j}(\sigma(t))(x_j - x_{k_0}) \right\rangle$$

$$\le \gamma \sum_{j \in \mathcal{N}_{k_0}(\sigma(t))} a_{k_0 j}(\sigma(t))\left(\varphi(x_j) - \varphi(x_{k_0})\right)$$

$$\le a^*\gamma(N-1)(d_\star + \varepsilon - \varphi(x_{k_0})). \tag{5.4}$$

It then follows that for all $t \ge \tilde{t}_{k_0}$,

$$\varphi(x_{k_0}(t)) \le e^{-\lambda_1(t - \tilde{t}_{k_0})}\varphi(x_{k_0}(\tilde{t}_{k_0})) + \left(1 - e^{-\lambda_1(t - \tilde{t}_{k_0})}\right)(d_\star + \varepsilon), \tag{5.5}$$

where $\lambda_1 = a^*\gamma(N-1)$.

Step 2.   Since $\mathcal{G}_{\sigma(t)}$ is uniformly jointly strongly connected, it is not hard to see that there exists an agent $k_1 \ne k_0 \in \mathcal{V}$ and $t_1 \ge \tilde{t}_{k_0}$ such that $(k_0, k_1) \in \mathcal{E}_{\sigma(t)}$ for $t \in [t_1, t_1 + \tau_D) \subseteq [\tilde{t}_{k_0}, \tilde{t}_{k_0} + T_0)$. Consider agent k_1 where $(k_0, k_1) \in \mathcal{E}_{\sigma(t)}$ for $t \in [\tilde{t}_{k_0}, \tilde{t}_{k_0} + T_0)$.

From (5.5), we obtain for all $t \in [\tilde{t}_{k_0}, \tilde{t}_{k_0} + (N-1)T_0]$,

$$\varphi(x_{k_0}(t)) \le \kappa_0 \triangleq \rho M_0 + (1-\rho)(d_\star + \varepsilon), \tag{5.6}$$

where $\rho = e^{-\lambda_1(N-1)T_0} = e^{-a^*\gamma(N-1)^2 T_0}$.

We next estimate $\varphi(x_{k_1}(t))$ by considering two different cases.

Case I: $\varphi(x_{k_1}(t)) > \varphi(x_{k_0}(t))$ for all $t \in [t_1, t_1 + \tau_D)$.

By using Assumption 5.1(ii), (5.1), (5.2), and (5.6), we obtain for all $t \in [t_1, t_1 + \tau_D)$,

$$\frac{d}{dt}\varphi(x_{k_1}(t)) \le \gamma \sum_{j \in \mathcal{N}_{k_1}(\sigma(t))\backslash\{k_0\}} a_{k_1 j}(\sigma(t))\left(\varphi(x_{k_j}) - \varphi(x_{k_1})\right)$$

$$+ a_{k_1 k_0}(\sigma(t))(\varphi(x_{k_0}) - \varphi(x_{k_1}))$$

$$\le a^*\gamma(N-2)(d_\star + \varepsilon - \varphi(x_{k_1})) + a_*\gamma\left(\kappa_0 - \varphi(x_{k_1})\right).$$

From the preceding relation, we obtain for $t \in [t_1, t_1 + \tau_D)$,

$$\varphi(x_{k_1}(t)) \le e^{-\lambda_2(t-t_1)}\varphi(x_{k_1}(t_1)) + \frac{\left[a^*(N-2)(d_\star+\varepsilon)+a_*\kappa_0\right](1-e^{-\lambda_2(t-t_1)})}{\lambda_2},$$

where $\lambda_2 = a^*\gamma(N-2) + a_*\gamma$. Therefore, we have

$$\varphi(x_{k_1}(t_1 + \tau_D)) \le \kappa_1 \triangleq \mu(d_\star + \varepsilon) + (1-\mu)\kappa_0, \tag{5.7}$$

where

$$\mu = \frac{\lambda_2 - a_*(1-e^{-\lambda_2\tau_D})}{\lambda_2}. \tag{5.8}$$

By applying the same analysis as we obtained (5.5) to the agent k_1, we obtain for all $t \ge t_1 + \tau_D$,

$$\varphi(x_{k_1}(t)) \le e^{-\lambda_1(t-(t_1+\tau_D))}\kappa_1 + \left[1 - e^{-\lambda_1(t-(t_1+\tau_D))}\right](d_\star + \varepsilon). \tag{5.9}$$

By combining the inequalities (5.6), (5.7), and (5.9), we obtain for all $t \in [t_1 + \tau_D, \tilde{t}_{k_0} + (N-1)T_0]$,

$$\varphi(x_{k_1}(t)) \le \varphi_1 M_0 + (1-\varphi_1)(d_\star + \varepsilon), \tag{5.10}$$

where $\varphi_1 = (1-\mu)\rho^2$.

Case II: There exists a time instant $\bar{t}_1 \in [t_1, t_1 + \tau_D)$ such that

$$\varphi(x_{k_1}(\bar{t}_1)) \le \varphi(x_{k_0}(\bar{t}_1)) \le \kappa_0. \tag{5.11}$$

By applying a similar analysis as we obtained (5.4) to the agent k_1, we obtain for all $t \ge \tilde{t}_{k_0}$,

$$\frac{d}{dt}\varphi(x_{k_1}(t)) \le a^*\gamma(N-1)(d_\star + \varepsilon - \varphi(x_{k_1}(t))).$$

This leads to

$$\varphi(x_{k_1}(t)) \le e^{-\lambda_1(t-\bar{t}_1)}\varphi(x_{k_1}(\bar{t}_1)) + (1 - e^{-\lambda_1(t-\bar{t}_1)})(d_\star + \varepsilon).$$

By combining the preceding relation, (5.6), and (5.11), and using $0 < \varphi_1 = (1 - \mu)\rho^2 < \rho^2$ which follows from $0 < \mu < 1$, we obtain for all $t \in [t_1 + \tau_D, \tilde{t}_{k_0} + (N - 1)T_0]$,

$$\varphi(x_{k_1}(t)) \le \rho^2 M_0 + (1 - \rho^2)(d_\star + \varepsilon) < \varphi_1 M_0 + (1 - \varphi_1)(d_\star + \varepsilon).$$

From the preceding relation and (5.10), it follows that for both cases, we have for all $t \in [t_1 + \tau_D, \tilde{t}_{k_0} + (N - 1)T_0]$,

$$\varphi(x_{k_1}(t)) \le \varphi_1 M_0 + (1 - \varphi_1)(d_\star + \varepsilon).$$

From the preceding relation, (5.6) and $0 < \varphi_1 < \rho < 1$, it follows that for all $t \in [t_1 + \tau_D, \tilde{t}_{k_0} + (N - 1)T_0]$,

$$\varphi(x_j(t)) \le \varphi_1 M_0 + (1 - \varphi_1)(d_\star + \varepsilon), \quad j \in \{k_0, k_1\}. \tag{5.12}$$

Step 3.   Consider agent $k_2 \notin \{k_0, k_1\}$ such that there exists an edge from the set $\{k_0, k_1\}$ to the agent k_2 in $\mathcal{E}_{\sigma(t)}$ for $t \in [t_2, t_2 + \tau_D) \subseteq [\tilde{t}_{k_0} + T_0, \tilde{t}_{k_0} + 2T_0)$. The existence of such an agent k_2 and t_2 follows similarly from the argument in Step 2.

Similarly, we can bound $\varphi(x_{k_2}(t))$ by considering two different cases and obtain that for all $t \in [t_2 + \tau_D, \tilde{t}_{k_0} + (N - 1)T_0]$,

$$\varphi(x_{k_2}(t)) \le \varphi_2 M_0 + (1 - \varphi_2)(d_\star + \varepsilon), \tag{5.13}$$

where $\varphi_2 = ((1 - \mu)\rho^2)^2$.

By combining (5.12) and (5.13), and using $0 < \varphi_2 < \varphi_1 < 1$, we obtain that for all $t \in [t_2 + \tau_D, \tilde{t}_{k_0} + (N - 1)T_0]$,

$$\varphi(x_j(t)) \le \varphi_2 M_0 + (1 - \varphi_2)(d_\star + \varepsilon), \quad j \in \{k_0, k_1, k_2\}.$$

Step 4.   By repeating the above process on time intervals $[\tilde{t}_{k_0} + 2T_0, \tilde{t}_{k_0} + 3T_0), \ldots, [\tilde{t}_{k_0} + (N - 2)T_0, \tilde{t}_{k_0} + (N - 1)T_0)$, we eventually obtain that for all $i \in \mathcal{V}$,

$$\varphi(x_i(\tilde{t}_{k_0} + (N - 1)T_0)) \le \varphi_{N-1} M_0 + (1 - \varphi_{N-1})(d_\star + \varepsilon),$$

where $\varphi_{N-1} = ((1 - \mu)\rho^2)^{N-1}$. The result of the lemma then follows by choosing $\bar{\rho} = \varphi_{N-1}$ and $\bar{t} = \tilde{t}_{k_0}$.

We can now use Lemma 5.3 to prove Theorem 5.2 by a contradiction argument. Suppose that there exists an agent $k_0 \in \mathcal{V}$ such that $0 \leq \alpha_{k_0} < d_\star$. It then follows from Lemma 5.3 that $\varphi(x_i(\bar{t} + (N-1)T_0)) < d_\star$ for all $i \in \mathcal{V}$, provided that $\varepsilon < \frac{\tilde{\rho}(d_\star - M_0)}{1-\tilde{\rho}}$. This contradicts the fact that $\lim_{t\to\infty} \max_{i\in\mathcal{V}} \varphi(x_i) = d_\star$. Thus, there does not exist an agent $k_0 \in \mathcal{V}$ such that $0 \leq \alpha_{k_0} < d_\star$. Hence, $\lim_{t\to\infty} \varphi(x_i(t)) = d_\star$ for all $i \in \mathcal{V}$.

5.3 Undirected Graph

In this section, we present the main results of the undirected graph with infinite joint connectivity.

Theorem 5.4 *Let Assumption 5.1 hold. Assume that $\mathcal{G}_{\sigma(t)}$ is undirected for all $t \geq 0$. The networked dynamical system (4.2) achieves global asymptotic φ-synchronization if $\mathcal{G}_{\sigma(t)}$ is infinitely jointly connected.*

Proof We first show that if there exists an agent $k_0 \in \mathcal{V}$ such that $0 \leq \alpha_{k_0} < d_\star$, there exist $0 < \tilde{\rho} < 1$ and $\tilde{t}$ such that

$$\varphi(x_i(\tilde{t} + \tau_D)) \leq \tilde{\rho} M_0 + (1 - \tilde{\rho})(d_\star + \varepsilon), \quad \forall i \in \mathcal{V}.$$

Step 1.   In this step, we focus on agent k_0. Since $\mathcal{G}_{\sigma(t)}$ is infinitely jointly connected, we can define

$$\hat{t}_1 \triangleq \inf_{t\in[\tilde{t}_{k_0},\infty)} \{\exists i \in \mathcal{V} \mid (k_0, i) \in \mathcal{E}_{\sigma(t)}\},$$

and the set

$$\mathcal{V}_1 \triangleq \{i \in \mathcal{V} \mid (k_0, i) \in \mathcal{E}_{\sigma(\hat{t}_1)}\}.$$

For $\tilde{t}_{k_0} \leq t < \hat{t}_1$, agent k_0 has no neighbor. It follows from Assumption 5.1(ii) that $\frac{d}{dt}\varphi(x_{k_0}(t)) = \langle \nabla\varphi(x_{k_0}), f(t, x_{k_0})\rangle \leq 0$. Thus, $\varphi(x_{k_0}(t)) \leq \varphi(x_{k_0}(\tilde{t}_{k_0}))$ for $\tilde{t}_{k_0} \leq t < \hat{t}_1$. By applying a similar analysis as we obtained (5.6), we have for all $t \in [\hat{t}_1, \hat{t}_1 + \tau_D)$,

$$\varphi(x_{k_0}(t)) \leq \hat{k}_0 \triangleq \hat{\rho} M_0 + (1 - \hat{\rho})(d_\star + \varepsilon),$$

where $\hat{\rho} = e^{-a^*(N-1)\tau_D}$.

Step 2.   In this step, we focus on all $k_1 \in \mathcal{V}_1$. We then estimate $\varphi(x_{k_1}(t))$ for all $k_1 \in \mathcal{V}_1$ by considering two different cases, i.e., Case I: If $\varphi(x_{k_1}(t)) > \varphi(x_{k_0}(t))$ for all $t \in [\hat{t}_1, \hat{t}_1 + \tau_D)$, and Case II: If there exists a time instant $\bar{t}_1 \in [\hat{t}_1, \hat{t}_1 + \tau_D)$ such that $\varphi(x_{k_1}(\bar{t}_1)) \leq \varphi(x_{k_0}(\bar{t}_1)) \leq \hat{k}_0$. By using a similar argument as the two-case analysis for agent k_1 in the proof of Theorem 5.2, we eventually obtain for all $j \in k_0 \cup \mathcal{V}_1$,

$$\varphi(x_j(\hat{t}_1 + \tau_D)) < \hat{\varphi}_1 M_0 + (1 - \hat{\varphi}_1)(d_\star + \varepsilon), \tag{5.14}$$

where $\hat{\varphi}_1 = (1 - \mu)\hat{\rho}^2$ with μ given by (5.8).

Step 3.   We then view the set $\{k_0\} \cup \mathcal{V}_1$ as a subsystem. Define $\hat{t}_2$ as the first time when there is an edge between this subsystem and the remaining agents and $\mathcal{V}_2$ accordingly. By using a similar analysis for agent k_1 in Step 2, we can estimate the upper bound for all agents in the set $\{k_0\} \cup \mathcal{V}_1 \cup \mathcal{V}_2$,

Step 4.   Since $\mathcal{G}_{\sigma(t)}$ is infinitely jointly connected, we can continue the above process until $\mathcal{V} = \{k_0\} \cup \mathcal{V}_1 \cup \cdots \cup \mathcal{V}_\ell$ for some $\ell \leq N - 1$. Eventually, we have

$$\varphi(x_i(\hat{t}_\ell + \tau_D)) < \hat{\varphi}_{N-1} M_0 + (1 - \hat{\varphi}_{N-1})(d_\star + \varepsilon), \quad \forall i \in \mathcal{V}.$$

By choosing $\tilde{t} = \hat{t}_\ell$ and $\tilde{\rho} = \hat{\varphi}_{N-1}$, we therefore have $\varphi(x_i(\tilde{t} + \tau_D)) \leq \tilde{\rho} M_0 + (1 - \tilde{\rho})(d_\star + \varepsilon), \quad \forall i \in \mathcal{V}$.

The remaining proof of Theorem 5.4 follows from a contradiction argument in the same way as the proof of Theorem 5.2.

Note that for the linear time-varying case $f(t, x) = A(t)x$, if there exists a matrix $P = P^T > 0$ such that

$$PA(t) + A^T(t)P \leq 0, \quad \forall t \geq 0, \tag{5.15}$$

it follows that $\varphi(x) = x^T Px$ for $x \in \mathbb{R}^n$ satisfies Assumption 5.1. In such a case, global asymptotic φ-synchronization is equivalent to global asymptotic P-norm synchronization (see Definition 4.5).

5.4   Simulations

In this section, we present two examples on φ-synchronization to illustrate our results.

5.4.1   Directed Graph

In the first example, we consider the multi-agent system (4.2) of four agents, where

$f(t, x_i) = Ax_i, x_i \in \mathbb{R}^3, i = 1, 2, 3, 4$. In particular, $A = \begin{bmatrix} \frac{1}{2} & \frac{1}{2} & -\frac{1}{2} \\ -\frac{1}{2} & -\frac{1}{2} & -\frac{1}{2} \\ 1 & 1 & 0 \end{bmatrix}$ and

$\gamma = 1$.

Fig. 5.1 φ-synchronization
for the directed graph case

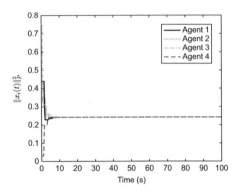

The switching topology $\mathcal{G}_{\sigma(t)}$ is selected the same as that of the directed
graph case in Sect. 4.5. Note that $\mathcal{G}([2s, 2s + 2))$ is uniformly jointly strongly
connected, and hence satisfies the condition in Theorem 5.2. Figure 5.1 shows that
global asymptotic φ-synchronization (P-norm synchronization) is achieved, where
$\varphi(x_i(t)) = \|x_i(t)\|_P^2$ with

$$P = \begin{bmatrix} 3 & 1 & -1 \\ 1 & 3 & 1 \\ -1 & 1 & 3 \end{bmatrix}. \tag{5.16}$$

5.4.2 Undirected Graph

Consider the multi-agent system (4.2) for the directed graph case, but let the com-
munications among the agents be undirected and have asymmetric edge weights. In
particular, the switching topology $\mathcal{G}_{\sigma(t)}$ is selected the same as that of the undirected
graph case in Sect. 4.5. Note that $\mathcal{G}_{\sigma(t)}$ is infinitely jointly connected, and hence
satisfies the condition in Theorem 5.4. Figure 5.2 shows that global asymptotic φ-
synchronization (P-norm synchronization) is achieved, where $\varphi(x_i(t)) = \|x_i(t)\|_P^2$
with P given by (5.16).

5.5 Literature

Synchronization of networked dynamical systems has been extensively studied in
recent years. The key idea of various distributed synchronization algorithms is
to realize a global emergence using only local information interactions [5, 12].
The synchronization problem of a single-integrator network has been fully studied
with an emphasis on the system robustness to the input time delays and switching

Fig. 5.2 φ-synchronization
for the undirected graph case

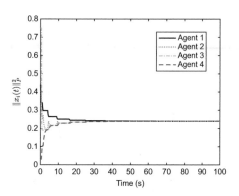

communication topologies [1, 5, 12, 15], discrete-time dynamical models [10, 28], nonlinear couplings [8], convergence speed [3], and leader–follower tracking [17].

Then, the researchers move their attention to the synchronization problem of multiple general linear dynamical systems. In particular, the authors of [23] generalize synchronization of multiple single-integrator systems to the case of multiple linear time-invariant high-order systems. For a network of neutrally stable systems and polynomially unstable systems, the author of [19] proposes a design scheme for achieving synchronization. The case of switching communication topologies is considered in [16] and a so-called consensus-based observer is proposed to guarantee leaderless synchronization of multiple identical linear dynamics systems under a jointly connected communication topology. Similar problems are also considered in [21] and [11], where a frequently connected communication topology is studied in [21] and an assumption on the neutral stability is imposed in [11]. The authors of [7] propose a neighbor-based observer to solve the synchronization problem for general linear time-invariant systems. In addition, the classical Laplacian matrix is generalized in [27] to a so-called interaction matrix, and a D-scaling approach is used to stabilize this interaction matrix. Synchronization of multiple heterogeneous linear systems has been investigated under both fixed and switching communication topologies [4, 9, 13, 24]. In [4], a high-gain approach is proposed to dominate the non-identical dynamics of the agents. The cases of frequently connected and jointly connected communication topologies are studied in [20] and [6], respectively, where a slow switching condition and a fast switching condition are presented.

We instead focus on synchronization of networked nonlinear dynamical systems with joint connectivity. The results in this chapter are based mainly on [25]. For further results on synchronization of networked nonlinear dynamical systems under time-varying communication topologies, see, e.g., [14, 22, 26, 29]. In particular, the result of this chapter has been extended to the leader–follower framework [26]. The authors of [29] focus on the case where the adjacency matrices associated with all communication graphs are simultaneously triangularizable. The authors of [14] consider switching communication graphs that are weakly connected and balanced at all times. A more general case where the switching communication graph frequently has a directed spanning tree has been considered in [22].

Acknowledgments ©2016 IEEE. Reprinted, with permission, from Tao Yang, Ziyang Meng, Guodong Shi, Yiguang Hong, Karl H. Johansson, "Network synchronization with nonlinear inherent dynamics and switching interactions", IEEE Transactions on Automatic Control. vol. 61, no. 10, pp. 3103–3108, 2016.

References

1. V.D. Blondel, J.M. Hendrickx, A. Olshevsky et al., Convergence in multiagent coordination, consensus, and flocking, in *Proceedings of the 44th IEEE Conference on Decision and Control*, Seville, Spain, 2005, pp. 2996–3000
2. S. Boyd, L. Vandenberghe, *Convex Optimization* (Cambridge University Press, Los Angeles, 2004)
3. M. Cao, A.S. Morse, B.D.O. Anderson, Reaching a consensus in a dynamically changing environment: convergence rates, measurement delays, and asynchronous events. SIAM J. Control Optim. **47**(2), 601–623 (2008)
4. H.F. Grip, T. Yang, A. Saberi et al., Output synchronization for heterogeneous networks of non-introspective agents. Automatica **48**(10), 2444–2453 (2012)
5. A. Jadbabaie, J. Lin, A.S. Morse, Coordination of groups of mobile autonomous agents using nearest neighbor rules. IEEE Trans. Autom. Control **48**(6), 988–1001 (2003)
6. H. Kim, H. Shim, J. Back et al., Consensus of output-coupled linear multi-agent systems under fast switching network: averaging approach. Automatica **49**(1), 267–272 (2013)
7. Z. Li, Z. Duan, G. Chen et al. Consensus of multiagent systems and synchronization of complex networks: a unified viewpoint. IEEE Trans. Circ. Syst. I Regul. Pap. **57**(1), 213–224 (2010)
8. Z. Lin, B. Francis, M. Maggiore, State agreement for continuous-time coupled nonlinear systems. SIAM J. Control Optim. **46**(1), 288–307 (2007)
9. J. Lunze, Synchronization of heterogeneous agents. IEEE Trans. Autom. Control **57**(11), 2885–2890 (2012)
10. L. Moreau, Stability of multi-agent systems with time-dependent communication links. IEEE Trans. Autom. Control **50**(2), 169–182 (2005)
11. W. Ni, D. Cheng, Leader-following consensus of multi-agent systems under fixed and switching topologies. Syst. Control Lett. **59**(3–4), 209–217 (2010)
12. R. Olfati-Saber, J.A. Fax, R.M. Murray, Consensus and cooperation in networked multi-agent systems. Proc. IEEE **95**(1), 215–233 (2007)
13. A. Pandey, Output consensus control for heterogeneous multi-agent systems, in *Proceedings of the 52nd IEEE Conference Decision and Control*, Florence, 2015, pp. 1502–1507
14. J. Qin, H. Gao, W.X. Zheng, Exponential synchronization of complex networks of linear systems and nonlinear oscillators: a unified analysis. IEEE Trans. Neur. Netw. Learn. Syst. **26**(3), 510–521 (2014)
15. W. Ren, R.W. Beard, Consensus seeking in multiagent systems under dynamically changing interaction topologies. IEEE Trans. Autom. Control **50**(5), 655–661 (2005)
16. L. Scardovi, R. Sepulchre, Synchronization in networks of identical linear systems. Automatica **45**, 546–551 (2009)
17. G. Shi, Y. Hong, K.H. Johansson, Connectivity and set tracking of multi-agent systems guided by multiple moving leaders. IEEE Trans. Autom. Control **57**(3), 663–676 (2012)
18. G. Shi, K.H. Johansson, Y. Hong, Reaching an optimal consensus: dynamical systems that compute intersections of convex sets. IEEE Trans. Autom. Control **58**(3), 610–622 (2013)
19. S.E. Tuna, Conditions for synchronizability in arrays of coupled linear systems. IEEE Trans. Autom. Control **54**(10), 2416–2420 (2009)
20. D. Vengertsev, H. Kim, H. Shim et al., Consensus of output-coupled linear multi-agent systems under frequently connected network, in *Proceedings of the IEEE Conference on Decision and Control*, Atlanta, 2010, pp. 4559–4564

21. J. Wang, D. Cheng, X. Hu, Consensus of multi-agent linear dynamic systems. Asian J. Control **10**(2), 144–155 (2008)

22. G. Wen, Z. Duan, G. Chen et al., Consensus tracking of multi-agent systems with Lipschitz-type node dynamics and switching topologies. IEEE Trans. Circ. Syst. I: Regul. Pap. **61**(2), 499–511 (2014)

23. P. Wieland, J.S. Kim, F. Allgöwer, On topology and dynamics of consensus among linear high-order agents. Int. J. Syst. Sci. **42**(10), 1831–1842 (2011)

24. P. Wieland, R. Sepulchre, F. Allgöwer, An internal model principle is necessary and sufficient for linear output synchronization. Automatica **47**(5), 1068–1074 (2011)

25. T. Yang, Z. Meng, G. Shi et al., Network synchronization with nonlinear dynamics and switching interactions. IEEE Trans. Autom. Control **61**(10), 3103–3108 (2016)

26. T. Yang, Z. Meng, G. Shi, Y. Hong, K.H. Johansson, Network synchronization with nonlinear dynamics and switching interactions (2014). Available at https://arxiv.org/abs/1401.6541

27. T. Yang, S. Roy, Y. Wan et al., Constructing consensus controllers for networks with identical general linear agents. Int. J. Rob. Nonlinear Control **21**(11), 1237–1256 (2011)

28. K. You, L. Xie, Network topology and communication data rate for consensusability of discrete-time multi-agent systems. IEEE Trans. Autom. Control **56**(10), 2262–2275 (2011)

29. J. Zhao, D.J. Hill, T. Liu, Passivity-based output synchronization of dynamical network with non-identical nodes, in *Proceedings of the IEEE Conference on Decision and Control*, Atlanta, 2010, pp. 7351–7356

Chapter 6
Periodic Solutions

In most of the networked dynamical system literature, the consensus and synchronization problems have been extensively studied. To illustrate the rich behaviors of networked dynamical systems, this chapter considers the existence of periodic behaviors for discrete-time second-order networked dynamical systems with input saturation constraints. We consider both cases where the agent dynamics are neutrally stable and double integrator, respectively. In both cases, we establish conditions on the feedback gains of the linear control law for achieving periodic behaviors. Simulation results are used to validate the theoretical results.

6.1 Problem Formulation

Periodic solutions are constructed to illustrate the rich behaviors of networked dynamical systems. To do so, we focus on the linear discrete-time systems with scalar inputs and specify (3.2) as

$$\bar{y}_i(k+1) = \bar{A}\bar{y}_i(k) + \bar{B}\sigma(u_i(k)), \quad i \in \mathcal{V}, \tag{6.1}$$

where $\bar{y}_i = [x_i; v_i] \in \mathbb{R}^2$, $u_i \in \mathbb{R}$, $\sigma(u_i)$ is the saturation function: $\sigma(u_i) = \operatorname{sgn}(u_i) \min\{1, |u_i|\}$, and the pair $(\bar{A}, \bar{B})$ describes the agent dynamics.

For networked dynamical system (6.1), we give the following definition on periodic solutions.

Definition 6.1 A solution $\bar{y}_i(k)$ of the networked dynamical system (6.1) is a periodic solution with period $\zeta > 0$, if for some initial states $\bar{y}_i(0)$, $i \in \mathcal{V}$, we have $\bar{y}_i(k + \zeta) = \bar{y}_i(k)$ for all $i \in \mathcal{V}$, and for all $k = 0, 1, \ldots$.

Consider a networked dynamical system of N identical discrete-time linear saturated model (6.1). The network among agents is described by an undirected

Z. Meng et al., *Modelling, Analysis, and Control of Networked Dynamical Systems*,
Systems & Control: Foundations & Applications,
https://doi.org/10.1007/978-3-030-84682-4_6

connected graph $\mathcal{G} = (\mathcal{V}, \mathcal{E})$, and for the weighted adjacency matrix $A = [a_{ij}] \in \mathbb{R}^{N \times N}$, we assume that $a_{ij} = a_{ji}$ for all $i, j \in \mathcal{V}$. The linear feedback control law with gain parameters α and β is given by:

$$u_i(k) = \sum_{j \in \mathcal{N}_i} a_{ij} \begin{bmatrix} \alpha & \beta \end{bmatrix} (\bar{y}_j(k) - \bar{y}_i(k)). \tag{6.2}$$

We next consider two cases where the agent dynamics are neutrally stable[1] and double integrator dynamics, respectively.

Before moving on, we need to define the following sets based on whether distance between agent $i \in \mathcal{V}$ and the root agent 1[2] is even or odd:

$$S_e = \{i \,|\, d(i, 1) = 0, 2, \ldots\}, \quad S_o = \{i \,|\, d(i, 1) = 1, 3, \ldots\}, \tag{6.3}$$

where the distance between two nodes i and j, $d(i, j)$ is the number of edges of a path between i and j minimized over all possible paths.

Let us also define

$$\bar{a} = \min_{\substack{(i,j) \in \mathcal{E} \\ i \in S_e, \, j \in S_o}} a_{ij}. \tag{6.4}$$

6.2 Neutrally Stable Agent Dynamics

We now consider the case where the agent dynamics is second-order neutrally stable and controllable agent dynamics, i.e., the pair $(\bar{A}, \bar{B})$ is in the following form:

$$\bar{A} = \begin{bmatrix} 0 & 1 \\ -1 & 2a \end{bmatrix}, \quad \bar{B} = \begin{bmatrix} 0 \\ 1 \end{bmatrix}, \tag{6.5}$$

where $-1 < a < 1$ and $a \neq 0$.

Our main result for this case is given below.

Theorem 6.2 *Consider the networked dynamical system* (6.1) *with the pair* $(\bar{A}, \bar{B})$ *given by* (6.5) *where* $-1 < a < 1$ *and* $a \neq 0$, *under the linear control law* (6.2). *Suppose that* $\mathcal{G}$ *is connected and the feedback gain parameters* α *and* β *satisfy*

$$|\alpha| \leq \operatorname{sign}(a)(\beta - \frac{a}{\bar{a}}), \tag{6.6}$$

---

[1] A discrete-time system is said to be neutrally stable if all its open-loop poles are within or on the unit circle with those on the unit circle being simple.

[2] Since the graph is connected, without loss of generality, we assume that agent 1 is the root agent.

where $\bar{a}$ is given by (6.4). Then there exist initial states such that the corresponding solution of the networked dynamical system is periodic with period $\zeta = 4$.

Proof A periodic solution with period $\zeta = 4$ is such that the input sequences (6.2) for all the agents are always in saturation. Moreover, it holds that

$$
\begin{cases}
u_i(k) \geq 1, & k = 0, 1, \\
u_i(k) \leq -1, & k = 2, 3,
\end{cases} \quad i \in S_e, \tag{6.7}
$$

$$
\begin{cases}
u_i(k) \leq -1, & k = 0, 1, \\
u_i(k) \geq 1, & k = 2, 3,
\end{cases} \quad i \in S_o. \tag{6.8}
$$

In what follows, we will show that (6.7) and (6.8) are satisfied for certain initial states $\bar{y}_i(0)$, $i \in \mathcal{V}$, and that the solution is periodic with period $\zeta = 4$. The proof is carried out in three steps.

Step 1: It follows from (6.1) and (6.7) that for $i \in S_e$, $\bar{y}_i(4) = \bar{A}^4 \bar{y}_i(0) + \bar{A}^3 \bar{B} + \bar{A}^2 \bar{B} - \bar{A}\bar{B} - \bar{B}$. Thus, in order to have $\bar{y}_i(0) = \bar{y}_i(4)$ for $i \in S_e$, we need

$$
\bar{y}_i(0) = (I - \bar{A}^4)^{-1}(\bar{A}^3 \bar{B} + \bar{A}^2 \bar{B} - \bar{A}\bar{B} - \bar{B}) = -(I + \bar{A}^2)^{-1}(I + \bar{A})\bar{B},
$$

where we have used the assumption on the eigenvalues of $\bar{A}$.
By plugging in the matrices $\bar{A}$ and $\bar{B}$ given in (6.5) into this equation, we obtain that

$$
\bar{y}_i(0) = \begin{bmatrix} \frac{1}{2a_i} \\ -\frac{1}{2a} \end{bmatrix} \text{ for } i \in S_e. \tag{6.9}
$$

Similarly, in order to have $\bar{y}_i(0) = \bar{y}_i(4)$ for $i \in S_o$, we need

$$
\bar{y}_i(0) = \begin{bmatrix} -\frac{1}{2a} \\ \frac{1}{2a} \end{bmatrix} \text{ for } i \in S_o. \tag{6.10}
$$

Step 2: In this step, we show that the four inequalities in either (6.7) or (6.8) can be reduced to two inequalities. For agent $j \in S_o$, we have

$$
u_j(k) = \sum_{i \in \mathcal{N}_j} a_{ij} \begin{bmatrix} \alpha & \beta \end{bmatrix} (\bar{y}_i(k) - \bar{y}_j(k))
$$

$$
= \sum_{i \in \mathcal{N}_j \cap S_e} a_{ij} \begin{bmatrix} \alpha & \beta \end{bmatrix} (\bar{y}_i(k) - \bar{y}_j(k)) + \sum_{i \in \mathcal{N}_j \cap S_o} a_{ij} \begin{bmatrix} \alpha & \beta \end{bmatrix} (\bar{y}_i(k) - \bar{y}_j(k)).
$$

Taking into account that $x_i(0) = x_j(0)$ and $v_i(0) = v_j(0)$ for all $i, j \in S_o$, we obtain

$$u_j(k) = \sum_{i \in \mathcal{N}_j \cap S_e} a_{ij} \begin{bmatrix} \alpha & \beta \end{bmatrix} (\bar{y}_i(k) - \bar{y}_j(k)). \tag{6.11}$$

From (6.11) and initial states given by (6.9) and (6.10), it is easy to verify that for $j \in S_o$, $u_j(k+2) \geq 1$ are equivalent to $u_j(k) \leq -1$ for $k = 0, 1$. Similarly, for $i \in S_e$, $u_i(k+2) \leq -1$ is equivalent to $u_i(k) \geq 1$ for $k = 0, 1$. Thus, the inequalities (6.7) and (6.8) are equivalent to the following inequalities: $u_i(0) \geq 1$ and $u_i(1) \geq 1$ for $i \in S_e$, and $u_j(0) \leq -1$ and $u_j(1) \leq -1$ for $j \in S_o$.

Step 3:   These inequalities are satisfied for each agent provided that for each edge $(i, j) \in \mathcal{E}$, where $i \in S_e$ and $j \in S_o$, the following two conditions:

$$a_{ij} \begin{bmatrix} \alpha & \beta \end{bmatrix} (\bar{y}_i(0) - \bar{y}_j(0)) = a_{ij} \frac{\alpha - \beta}{a} \leq -1,$$

$$a_{ij} \begin{bmatrix} \alpha & \beta \end{bmatrix} (\bar{y}_i(1) - \bar{y}_j(1)) = a_{ij} \frac{-\alpha - \beta}{a} \leq -1,$$

are satisfied. It is easy to see that this is the case, if the feedback gain parameters α and β satisfy (6.6).

From the above analysis, it follows that the solution of the networked dynamical system is periodic with period $\zeta = 4$, for initial states satisfying (6.9) and (6.10).

6.3   Double Integrator Agent Dynamics

In this section, we consider the case where the agent dynamics is double integrator dynamics, i.e., $\bar{A}$ and $\bar{B}$ in (6.1) are given by

$$\bar{A} = \begin{bmatrix} 1 & 1 \\ 0 & 1 \end{bmatrix}, \quad \bar{B} = \begin{bmatrix} 0 \\ 1 \end{bmatrix}. \tag{6.12}$$

Our main result for this case is given below.

Theorem 6.3 *Consider the networked dynamical system (6.1) with the pair $(\bar{A}, \bar{B})$ given by (6.12) under the linear control law (6.2). Suppose that $\mathcal{G}$ is connected and the feedback gain parameters α and β satisfy*

$$0 < \alpha < \beta < \tfrac{3}{2}\alpha. \tag{6.13}$$

Then there exists initial states such that the corresponding solution of the networked dynamical system is periodic with the period $\zeta = 2m$, where

$$m \geq \frac{4(\alpha - \beta) + \frac{2}{a}}{3\alpha - 2\beta}. \tag{6.14}$$

Proof We prove this theorem by explicitly constructing periodic solutions with an even period $\zeta = 2m$. The periodic solution that we will construct is such that the input sequences (6.2) for all the agents are always in saturation, and for $i \in S_e$,

$$u_i(k) \geq 1, k = 0, \ldots, m - 1, \ u_i(k) \leq -1, k = m, \ldots, 2m - 1, \tag{6.15}$$

and for $i \in S_o$,

$$u_i(k) \leq -1, k = 0, \ldots, m - 1, \ u_i(k) \geq 1, k = m, \ldots, 2m - 1. \tag{6.16}$$

In what follows, we will show that (6.15) and (6.16) are satisfied for certain m and initial states $x_i(0)$ and $v_i(0)$ for $i \in V$, and that the solution is periodic with period $\zeta = 2m$ for these initial states. The proof has three steps.

Step 1:   It is sufficient to show that $x_i(\zeta) = x_i(0)$ and $v_i(\zeta) = v_i(0)$ for all $i \in V$. It follows from (6.1), (6.15), and (6.16) that $v_i(\zeta) = v_i(0)$ for all $i \in V$. It is also easy to obtain that $x_i(2m) = x_i(0) + 2mv_i(0) + m^2$ for $i \in S_e$ and $x_i(2m) = x_i(0) + 2mv_i(0) - m^2$ for $i \in S_o$. Thus, in order to have $x_i(\zeta) = x_i(0)$ for all $i \in V$, we must have that

$$\begin{cases} v_i(0) = -m/2, & i \in S_e, \\ v_i(0) = m/2, & i \in S_o. \end{cases} \tag{6.17}$$

Step 2:   We show that the $2m$ inequalities in either (6.15) or (6.16) can be reduced to two inequalities by appropriately choosing initial states $x_i(0)$ for some $i \in V$.

Step 2.1:   For agent $j \in S_o$, we have

$$u_j(k) = \sum_{i \in \mathcal{N}_j} a_{ij} \begin{bmatrix} \alpha & \beta \end{bmatrix} (\bar{y}_i(k) - \bar{y}_j(k))$$

$$= \sum_{i \in \mathcal{N}_j \cap S_e} a_{ij} \begin{bmatrix} \alpha & \beta \end{bmatrix} (\bar{y}_i(k) - \bar{y}_j(k)) + \sum_{i \in \mathcal{N}_j \cap S_o} a_{ij} \begin{bmatrix} \alpha & \beta \end{bmatrix} (\bar{y}_i(k) - \bar{y}_j(k)).$$

Choosing $x_i(0) = x_j(0)$ for $i \in S_o$ if $(i, j) \in \mathcal{E}$, and using $v_i(0) = v_j(0)$ for all $i, j \in S_o$, gives

$$u_j(k) = \sum_{i \in \mathcal{N}_j \cap S_e} a_{ij} \begin{bmatrix} \alpha & \beta \end{bmatrix} (\bar{y}_i(k) - \bar{y}_j(k)). \tag{6.18}$$

Similarly, for agent $i \in S_e$, choosing $x_j(0) = x_i(0)$ for $j \in S_e$ if $(i, j) \in \mathcal{E}$, yields

$$u_i(k) = \sum_{j \in \mathcal{N}_i \cap S_o} a_{ij} \begin{bmatrix} \alpha & \beta \end{bmatrix} (\bar{y}_i(k) - \bar{y}_j(k)).$$

Step 2.2:   Let us now focus on any edge $(i, j) \in \mathcal{E}$, such that $i \in S_e$ and $j \in S_o$. We first note that $0 < \alpha < \beta$ from (6.13) implies that $\beta > \alpha - \frac{1}{2}k\alpha$ for $k = 0, \ldots, m - 1$, which yields $-\frac{\alpha m}{2} + \beta > \frac{1}{2}\alpha(-m - k + 2)$. Since $m - k - 1 \geq 0$, multiplying the above inequality on both sides with $m - k - 1$ yields

$$-\tfrac{\alpha m}{2}(m - k - 1) + \beta(m - k - 1) \geq \alpha[\tfrac{k(k-1)}{2} - \tfrac{(m-1)(m-2)}{2}].$$

This is equivalent to that

$$a_{ij}\begin{bmatrix} \alpha & \beta \end{bmatrix}(\bar{y}_i(m - 1) - \bar{y}_j(m - 1)) \geq a_{ij}\begin{bmatrix} \alpha & \beta \end{bmatrix}(\bar{y}_i(k) - \bar{y}_j(k)) \qquad (6.19)$$

for $k = 0, \ldots, m - 1$, since $v_i(0) = -\frac{m}{2}$ for $i \in S_e$, $v_j(0) = \frac{m}{2}$ for $j \in S_o$, and $a_{ij} \geq 0$.

Step 2.3:   Since the inequality (6.19) holds for each $i \in \mathcal{N}_j \cap S_e$, where $j \in S_o$, adding them up together with (6.18) yields that

$$u_j(m - 1) \geq u_j(k), \quad k = 0, \ldots, m - 1, \quad j \in S_o.$$

Hence, for $j \in S_o$, $u_j(m - 1) \leq -1$ implies that $u_j(k) \leq -1$ for all $k = 0, \ldots, m - 1$.

A similar argument shows that

$$u_j(2m - 1) \leq u_j(k), \quad k = m, \ldots, 2m - 1, \quad j \in S_o.$$

Hence, for $j \in S_o$, $u_j(2m - 1) \geq 1$ implies that $u_j(k) \geq 1$ for all $k = m, \ldots, 2m - 1$.

Similarly, we can show that for $i \in S_e$, $u_i(m-1) \geq 1$ implies that $u_i(k) \geq 1$ for $k = 0, \ldots, m - 1$, and that $u_i(2m - 1) \leq -1$ implies that $u_i(k) \leq -1$ for $k = m, \ldots, 2m - 1$.

To summarize, if there is an edge connecting agents within S_e or S_o, we set their initial states the same, i.e.,

$$x_i(0) = x_j(0) \text{ for } (i, j) \in \mathcal{E}, \text{ if } i, j \in S_e, \text{ or } i, j \in S_o. \qquad (6.20)$$

Then inequalities (6.15) or (6.16) are reduced to $u_j(m-1) \leq -1$ and $u_j(2m - 1) \geq 1$ for $j \in S_o$, or $u_i(m - 1) \geq 1$ and $u_i(2m - 1) \leq -1$ for $i \in S_e$.

Step 3:   It is clear that these two inequalities are satisfied provided that for each edge $(i, j) \in \mathcal{E}$, where $i \in S_e$ and $j \in S_o$, the following two conditions:

$$a_{ij}\begin{bmatrix} \alpha & \beta \end{bmatrix}(\bar{y}_i(m - 1) - \bar{y}_j(m - 1))$$

$$= a_{ij}\left\{\alpha\begin{bmatrix} x_i(0) - x_j(0) - 2m + 2 \end{bmatrix} + \beta(m - 2)\right\} \leq -1,$$

$$a_{ij} \begin{bmatrix} \alpha & \beta \end{bmatrix} (\bar{y}_i(2m-1) - \bar{y}_j(2m-1))$$

$$= a_{ij} \left\{ \alpha \left[x_i(0) - x_j(0) + m - 2 \right] - \beta(m-2) \right\} \geq 1,$$

are satisfied. They are equivalent to

$$\frac{1}{a_{ij}} + (\beta - \alpha)(m-2) \leq \alpha(x_i(0) - x_j(0)) \leq 2\alpha(m-1) - \beta(m-2) - \frac{1}{a_{ij}}. \quad (6.21)$$

We see that suitable $x_i(0)$ and $x_j(0)$, where $i \in S_e$, $j \in S_o$, and $(i, j) \in \mathcal{E}$, exist
if

$$\frac{1}{a_{ij}} + (\beta - \alpha)(m-2) \leq 2\alpha(m-1) - \beta(m-2) - \frac{1}{a_{ij}}. \quad (6.22)$$

For $m > 2$, (6.22) is equivalent to $\beta \leq \frac{3m-4}{2m-4}\alpha - \frac{1}{a_{ij}(m-2)}$. If we take the value of
m to be sufficiently large, we obtain that

$$\beta \leq \lim_{m \to +\infty} \left[\frac{3m-4}{2m-4}\alpha - \frac{1}{a_{ij}(m-2)} \right] = \frac{3}{2}\alpha.$$

Therefore for any α and β which satisfy (6.13), if condition (6.14) holds, then
(6.22) is satisfied.

From the above analysis, we see that the solution of the networked dynamical
system is periodic with period $\zeta = 2m$, where m satisfies (6.14), for initial states
satisfying (6.17), (6.20), and (6.21).

6.4 Simulations

In this section, we present several examples to illustrate the results. The network
consists of $N = 7$ agents, and the topology is given by the undirected weighted
graph depicted in Fig. 6.1.

Fig. 6.1 Network with seven
agents

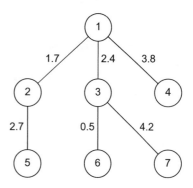

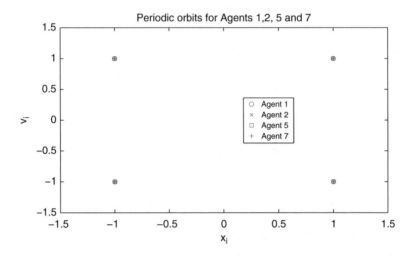

Fig. 6.2 Periodic behavior of period 4 for the neutrally stable dynamics

6.4.1 Neutrally Stable Agent Dynamics

We first consider the neutrally stable case, that is, the matrices $\bar{A}$ and $\bar{B}$ in (6.1) are given by (6.5) with $a = \frac{1}{2}$, which results in eigenvalues $\frac{1}{2} \pm \frac{\sqrt{3}}{2}\mathbf{j}$ on the unit circle. It is easy to see that $\bar{a} = a_{36} = 0.5$. Thus, (6.6) becomes $-\beta + 1 \leq \alpha \leq \beta - 1$. We then choose the feedback gain parameters $\alpha = 0.5$ and $\beta = 2$. Figure 6.2 shows that the networked dynamical system exhibits a periodic behavior with period $\zeta = 4$ for initial states $x_i(0) = 1$, $v_i(0) = -1$ for $i \in S_e = \{1, 5, 6, 7\}$ and $x_i(0) = -1$, $v_i(0) = 1$ for $i \in S_o = \{2, 3, 4\}$, which satisfy (6.9) and (6.10).

6.4.2 Double Integrator Agent Dynamics

We next consider the double integrator case, that is, the matrices $\bar{A}$ and $\bar{B}$ in (6.1) are given by (6.12). Then, we choose the feedback gain parameters $\alpha = 1$ and $\beta = 1.2$. Therefore the sufficient condition for achieving a periodic behavior (6.13) is satisfied. It is easy to see that $\bar{a} = a_{36} = 0.5$. Therefore we choose $m = 11$ so that (6.14) is satisfied. From the proof of Theorem 6.3, we see that the networked dynamical system exhibits a periodic solution with $\zeta = 22$ if the initial states satisfy (6.17) and (6.21) with $m = 11$, i.e., $v_i(0) = -5.5$ for $i \in S_e = \{1, 5, 6, 7\}$, $v_i(0) = 5.5$ for $i \in S_o = \{2, 3, 4\}$, $2.3882 \leq x_1(0) - x_2(0) \leq 8.6118$, $2.2167 \leq x_1(0) - x_3(0) \leq 8.7833$, $2.0632 \leq x_1(0) - x_4(0) \leq 8.9368$, $2.1704 \leq x_5(0) - x_2(0) \leq 8.8296$, $3.8000 \leq x_6(0) - x_3(0) \leq 7.2000$, and $2.0381 \leq x_7(0) - x_3(0) \leq 8.9619$. We then choose $x_1(0) = 21$, $x_2(0) = 16$, $x_3(0) = 17$, $x_4(0) = 15$, $x_5(0) = 23$, $x_6(0) = 22$, and $x_7(0) = 24$ so that the above conditions are satisfied. With

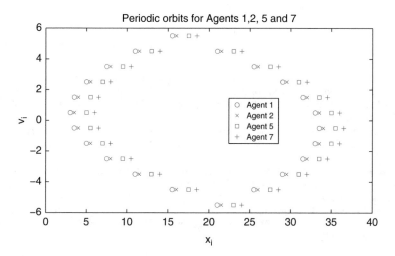

Fig. 6.3 Periodic solutions of period 22 for the double integrator case

these initial states, the networked dynamical system with input saturation constraints exhibits a periodic behavior of period $\zeta = 2m = 22$ as shown in Fig. 6.3, where state trajectories for agents 1, 2, 5, and 7 are given.

6.5 Literature

The results in this chapter are based mainly on [6]. Some related results can be found in [1–5]. In particular, the authors of [1] show the existence of the periodic phenomenon for discrete-time single-integrator networked dynamical systems, while the double integrator agent dynamics has been considered in [3]. In this chapter, we have shown the existence of general second-order networked dynamical systems including both neutrally stable and double integrator dynamics. Such an extension is not only challenging since the dynamics may diverge without control laws, but also useful since many physical systems can be modeled as such systems. Moreover, this chapter also extends the existence of periodic behavior for individual discrete-time system [2, 4, 5], to networked dynamical systems. Although this chapter only studies the second-order networked dynamical systems, these systems are known as a key benchmark for dynamical behavior of nonlinear networked dynamical systems. By fully understanding these systems, we make a key step in understanding the abilities of linear consensus control law for achieving periodic behaviors.

Acknowledgments ©2016 IEEE. Reprinted, with permission, from Tao Yang, Ziyang Meng, Dimos V. Dimarogonas, Karl H. Johansson, "Periodic behaviors for discrete-Time second-order multiagent systems with input saturation constraints", IEEE Transactions on Circuits and Systems II: Berif Papers, vol. 63, no. 7, pp. 663–667, 2016.

References

1. Y. Chen, J. Lü, F. Han et al. On the cluster consensus of discrete-time multi-agent systems. Syst. Control Lett. **60**(7), 517–523 (2011)
2. W. Rugh, *Linear Systems Theory* (Prentice Hall, Upper Saddle River, 1993)
3. T. Yang, Z. Meng, D.V. Dimarogonas et al., Periodic behaviors in multi-agent systems with input saturation constraints, in *Proceedings of the IEEE Conference on Decision and Control*, Firenze, 2013, pp. 4467–4472
4. T. Yang, A.A. Stoorvogel, A. Saberi, Dynamic behavior of the discrete-time double integrator with saturated locally stabilizing linear state feedback laws. Int. J. Robust Nonlinear Control **23**(17), 1899–1931 (2013)
5. T. Yang, A.A. Stoorvogel, A. Saberi et al., Further results on saturated globally stabilizing linear state feedback control laws for single-input neutrally stable planar systems, in *Proceedings of the European Control Conference*, Zürich, 2013, pp. 2728–2733
6. T. Yang, Z. Meng, D.V. Dimarogonas et al., Periodic behaviors for discrete-time second-order multiagent systems with input saturation constraints. IEEE Trans. Circ. Syst. II: Express Briefs **63**(7), 663–667 (2016)

Chapter 7
Modulus Consensus

Motivated by the opinion dynamics evolving over social networks [8, 23], behavior analysis of the state consensus problem with both cooperative and antagonistic interactions attracts much attention recently. By an antagonistic interaction between a pair of nodes updating their states we mean that one node receives the opposite of the state of the other, and naturally by a cooperative interaction we mean that the former receives the true state of the latter.

In this chapter, we study the discrete-time consensus problem over networks with antagonistic and cooperative interactions. The pairwise communication can be either directed or undirected, and the overall communication topology may change with time. The concept of modulus consensus is introduced to characterize the scenario that the moduli of the node states reach consensus. It is proven that modulus consensus is achieved if the switching interaction graph is uniformly jointly strongly connected for directed graphs, or infinitely jointly connected for undirected graphs. Finally, simulation results using a discrete-time Kuramoto model are given to illustrate the convergence results showing that the proposed framework is applicable to a class of networks with general nonlinear node dynamics.

7.1 Definitions on Modulus Consensus

In the network with both cooperative and antagonistic interactions, the interaction graph of the network needs to be redefined. Consider $\mathcal{G}_k = (\mathcal{V}, \mathcal{E}_k)$, $k = 0, 1, \ldots$, with node set $\mathcal{V}$, and $\mathcal{E}_k \subseteq \mathcal{V} \times \mathcal{V}$ is the set of edges at time k. We assume that $\mathcal{G}_k$ is a signed graph, where "+" or "−" is associated with each edge $(i, j) \in \mathcal{E}_k$. Here, "+" represents cooperative relation, and "−" represents antagonistic relation. The set of neighbors of node i in $\mathcal{G}_k$ is denoted by $\mathcal{N}_i(k) := \{j : (j, i) \in \mathcal{E}_k\}$, and $\mathcal{N}_i^+(k)$ and $\mathcal{N}_i^-(k)$ are used to denote the cooperative neighbor sets and antagonistic neighbor

© The Author(s), under exclusive license to Springer Nature Switzerland AG 2021
Z. Meng et al., *Modelling, Analysis, and Control of Networked Dynamical Systems*,
Systems & Control: Foundations & Applications,
https://doi.org/10.1007/978-3-030-84682-4_7

sets, respectively. Clearly, $\mathcal{N}_i(k) = \mathcal{N}_i^-(k) \cup \mathcal{N}_i^+(k)$. The joint graph of $\mathcal{G}$ during time interval $[k_1, k_2)$ is defined by $\mathcal{G}([k_1, k_2)) = \bigcup_{k \in [k_1,k_2)} \mathcal{G}_k = (\mathcal{V}, \bigcup_{k \in [k_1,k_2)} \mathcal{E}_k)$.

For the agent dynamics, we use the simplified version of discrete-time model (3.2):

$$x_i(k+1) = \sum_{j \in \mathcal{N}_i(k) \cup \{i\}} a_{ij}(x, k) x_j(k), \quad k = 0, 1, \ldots, \ i \in \mathcal{V}, \tag{7.1}$$

where $x_i(k) \in \mathbb{R}^n$ represents of state of agent i at time k, $x = [x_1^T, x_2^T, \ldots, x_N^T]^T$, and $a_{ij}(x, k)$ represents the weight of edge (j, i). We next present the definitions on asymptotic modulus consensus.

Definition 7.1 System (7.1) achieves asymptotic modulus consensus for any initial state $x(0) \in \mathbb{R}^{Nn}$ if

$$\lim_{k \to \infty} \|x(k)\|_{\mathcal{J}} = 0,$$

where $x = [x_1^T, x_2^T, \ldots, x_N^T]^T$, $y = [y_1^T, y_2^T, \ldots, y_N^T]^T$, and $\mathcal{J} = \{y \in \mathbb{R}^{Nn} : |y_{1,q}| = |y_{2,q}| = \cdots = |y_{N,q}|, \ \forall q\}$, $y_{i,q}$ denotes the qth entry of y_i, and $\|x\|_{\mathcal{J}} = \inf_{y \in \mathcal{J}} \|x - y\|$.

7.2   Problem Formulation

Without loss of generality, consider N agents and the state space for the agents is $\mathbb{R}$. In addition, let $x_i \in \mathbb{R}$ be subject to the model (7.1) and denote $x = (x_1, x_2, \ldots, x_N)^T$.

We know that (7.1) can be written in the compact form:

$$x(k+1) = A(x, k) x(k), \quad k = 0, 1, \ldots, \tag{7.2}$$

where $A(x, k) = [a_{ij}(x, k)] \in \mathbb{R}^{N \times N}$ denotes signed weight matrix. For $a_{ij}(x, k)$, we impose the following standing assumption.

Assumption 7.1 *There exists a positive constant $0 < \lambda < 1$ such that:*

(i) $a_{ii}(x, k) \geq \lambda$, *for all i, x, k, and $\sum_{j \in \mathcal{N}_i(k) \cup \{i\}} |a_{ij}(x, k)| = 1$ for all i, x, k;*

(ii) $a_{ij}(x, k) \geq \lambda$ *for all i, j, x, k, if $j \in \mathcal{N}_i^+(k)$; and $a_{ij}(x, k) \leq -\lambda$ for all i, j, x, k, if $j \in \mathcal{N}_i^-(k)$.*

Clearly under our standing assumption, this model describes the corresponding discrete-time cooperative-antagonistic interactions introduced by [2] in the sense that $a_{ij}(x, k) > 0$ represents that i is cooperative to j, and $a_{ij}(x, k) < 0$ represents that i is antagonistic to j.

We next compare the concepts of "consensus" [17] and "bipartite consensus" [2] with the proposed "modulus consensus." Consensus means that states of the agents converge to the same value, i.e., $\lim_{k\to\infty} x_i(k) = \alpha_1$ for all $i \in \mathcal{V}$. Bipartite consensus means that the absolute value of states of the agents converges a nonzero same state, i.e., $\lim_{k\to\infty} |x_i(k)| = \alpha_2 > 0$ for all $i \in \mathcal{V}$. In contrast, modulus consensus proposed in this chapter allows that different agents converge to zero state, a nonzero same state, or split into two different states.

7.3   Modulus Consensus Conditions

The main result for general directed graphs is first presented indicating that uniform joint strong connectivity is sufficient for modulus consensus. Before moving on, a key technical lemma is first established. Since the proofs do not rely on whether or not a_{ij} depends on x, without loss of generality, we use $a_{ij}(k)$ to denote $a_{ij}(x, k)$.

Lemma 7.2 *Suppose that Assumption 7.1 holds. For system (7.1), it holds that* $\|x(k+1)\|_\infty \le \|x(k)\|_\infty$, *for all* $k = 0, 1, \dots.$

Proof It follows from Assumption 7.1 that $|x_i(k+1)| \le \sum_{j\in\mathcal{N}_i(k)\cup\{i\}} |a_{ij}(k)|$ $\times |x_j(k)| \le \left(\sum_{j\in\mathcal{N}_i(k)\cup\{i\}} |a_{ij}(k)|\right) \max_{i\in\mathcal{V}} |x_i(k)| = \|x(k)\|_\infty$, for all $i \in \mathcal{V}$, which leads to the conclusion directly.

Theorem 7.3 *Suppose that Assumption 7.1 holds. System (7.1) achieves asymptotic modulus consensus for all initial states $x(0) \in \mathbb{R}^N$ if $\{\mathcal{G}_k\}_0^\infty$ is uniformly jointly strongly connected (see Definition 3.1).*

Proof We first show that consensus is achieved in absolute values, i.e., $\lim_{k\to\infty} |x_i(k)| = M^*$, for all $i \in \mathcal{V}$. In particular, since a bounded monotone sequence always admits a limit, Lemma 7.2 implies that for any initial states $x(0)$, there exists a constant M^*, such that $\lim_{k\to\infty} \|x(k)\|_\infty = M^*$. We further define $\Phi_i = \limsup_{k\to\infty} |x_i(k)|$, $\Psi_i = \liminf_{k\to\infty} |x_i(k)|$, for all $i \in \mathcal{V}$. Clearly, it must hold that $0 \le \Psi_i \le \Phi_i \le M^*$. Therefore, $\lim_{k\to\infty} |x_i(k)| = M^*$, for all $i \in \mathcal{V}$ if and only if $\Phi_i = \Psi_i = M^*$, $i \in \mathcal{V}$.

In addition, based on the fact that $\lim_{k\to\infty} \|x(k)\|_\infty = M^*$, it follows that for any $\varepsilon > 0$, there exists a $\widehat{k}(\varepsilon) > 0$ such that $M^* - \varepsilon \le \|x(k)\|_\infty \le M^* + \varepsilon$, $\forall k \ge \widehat{k}(\varepsilon)$ and $|x_i(k)| \le M^* + \varepsilon$, $\forall i \in \mathcal{V}$, $\forall k \ge \widehat{k}(\varepsilon)$.

We next use a contradiction argument. Now suppose that there exists a node $i_1 \in \mathcal{V}$ such that $0 \le \Psi_{i_1} < M^*$. With the definitions of Ψ_{i_1}, for any $\varepsilon > 0$, there exist a constant $\Psi_{i_1} < \alpha_1 < M^*$ and a time instance $k_1 \ge \widehat{k}(\varepsilon)$ such that $|x_{i_1}(k_1)| \le \alpha_1$. This shows that

$$|x_{i_1}(k_1)| \le M^* - (M^* - \alpha_1) = M^* - \xi_1, \tag{7.3}$$

where $\xi_1 = M^* - \alpha_1 > 0$.

First of all, it follows from Lemma 7.2 that $|x_j(k_1 + s)| \leq \|x(k_1)\|_\infty$, for all $s = 1, 2, \ldots$ and all $j \in \mathcal{V}$. We next fix i_1 and analyze the trajectory of x_{i_1} after k_1. Then it must be true that for all $s = 1, 2, \ldots$,

$$
\begin{aligned}
|x_{i_1}(k_1 + s)| \leq & \sum_{j \in \mathcal{N}_{i_1}(k_1+s-1)\cup\{i_1\}} |a_{i_1 j}(k_1 + s - 1)| \, |x_j(k_1 + s - 1)| \\
= & \, |a_{i_1 i_1}(k_1 + s - 1)| \, |x_{i_1}(k_1 + s - 1)| \\
& + \sum_{j \in \mathcal{N}_{i_1}(k_1+s-1)} |a_{i_1 j}(k_1 + s - 1)| \|x_j(k_1 + s - 1)| \\
\leq & \, |a_{i_1 i_1}(k_1 + s - 1)| \|x_{i_1}(k_1 + s - 1)| \\
& + (1 - |a_{i_1 i_1}(k_1 + s - 1)|) \|x(k_1)\|_\infty.
\end{aligned}
$$

Also note that $\|x(k_1)\|_\infty \leq M^* + \varepsilon$. It thus follows from (7.3) that

$$
\begin{aligned}
|x_{i_1}(k_1 + 1)| &\leq |a_{i_1 i_1}(k_1)| |x_{i_1}(k_1)| + (1 - |a_{i_1 i_1}(k_1)|) \|x(k_1)\|_\infty \\
&\leq a_{i_1 i_1}(k_1)(M^* - \xi_1) + (1 - a_{i_1 i_1}(k_1))(M^* + \varepsilon) \\
&\leq M^* + \varepsilon - \lambda \xi_1,
\end{aligned}
\tag{7.4}
$$

where we have used the fact that $a_{i_1 i_1} \geq \lambda$ from Assumption 7.1.

By a recursive analysis, we can further deduce that

$$
|x_{i_1}(k_1 + s)| \leq M^* + \varepsilon - \lambda^s \xi_1, \quad s = 1, 2, \ldots.
\tag{7.5}
$$

Next, we consider the time interval $[k_1, k_1 + T)$. Since $\mathcal{G}([k_1, k_1 + T))$ is strongly connected, there is a path from i_1 to any other node during the time interval $[k_1, k_1 + T)$. This implies that there exists a time instant $k_2 \in [k_1, k_1 + T)$ such that i_1 is a neighbor of another node i_2 at k_2. We next analyze the trajectory of x_{i_2} after k_2. It follows that $|x_{i_2}(k_2 + s)| \leq \sum_{j \in \mathcal{N}_{i_2}(k_2+s-1)\cup\{i_2\}} |a_{i_2 j}(k_2 + s - 1)| |x_j(k_2 + s - 1)| = |a_{i_2 i_1}(k_2 + s - 1)| |x_{i_1}(k_2 + s - 1)| + \sum_{j \in \mathcal{N}_{i_2}(k_2+s-1)\cup\{i_2\}\setminus\{i_1\}} |a_{i_2 j}(k_2 + s - 1)| |x_j(k_2 + s - 1)| \leq |a_{i_2 i_1}(k_2 + s - 1)| |x_{i_1}(k_2 + s - 1)| + (1 - |a_{i_2 i_1}(k_2 + s - 1)|) \|x(k_2)\|_\infty$, where we have used the fact that $|x_i(k_2 + s)| \leq \|x(k_2)\|_\infty$, for all $s = 1, 2, \ldots$, and for all $i \in \mathcal{V}$. Noting that $\|x(k_2)\|_\infty \leq M^* + \varepsilon$, it thus follows that $|x_{i_2}(k_2 + s)| \leq |a_{i_2 i_1}(k_2 + s - 1)|(M^* + \varepsilon - \lambda^{s-1+k_2-k_1}\xi_1) + (1 - |a_{i_2 i_1}(k_2 + s - 1)|)(M^* + \varepsilon) \leq M^* + \varepsilon - \lambda^{s+k_2-k_1}\xi_1$, $s = 1, 2, \ldots$. We can further use the fact $k_2 - k_1 < T$ to obtain

$$
|x_{i_2}(k_1 + s)| \leq M^* + \varepsilon - \lambda^s \xi_1, \quad s = T, T + 1, \ldots.
\tag{7.6}
$$

We now reiterate the previous argument for the time interval $[k_1 + T, k_1 + 2T)$. Again, there is a path from i_1 to any other node during the time interval $[k_1 + T, k_1 + 2T)$. There exists a time instant $k_3 \in [k_1 + T + 1, k_1 + 2T)$ such that either i_1 or

i_2 is a neighbor of i_3 (i_3 is another node different from i_1 and i_2) at k_3. For any of the two cases we can deduce from (7.5) and (7.6) that for agent i_3, it must hold $|x_{i_3}(k_1 + s)| \leq M^* + \varepsilon - \lambda^s \xi_1$, $s = 2T, 2T + 1, \ldots$.

The above analysis can be carried out to intervals $[k_1 + 2T, k_1 + 3T), \ldots, [k_1 + (n-2)T, k_1 + (N-1)T)$, where $i_4, \ldots, i_n$ can be found recursively until they include the whole network. We can therefore finally arrive at $\|x(k_1 + (N-1)T)\|_\infty \leq M^* + \varepsilon - \lambda^{(N-1)T}\xi_1 < M^* - \lambda^{(N-1)T}\xi_1/2$, for sufficient small ε satisfying $\varepsilon < \lambda^{(N-1)T}\xi_1/2$. Then, it follows from Lemma 7.2 that

$$\|x(k)\|_\infty < M^* - \lambda^{(N-1)T}\xi_1/2,$$

for all $k \geq k_1 + (N-1)T$, which contradicts the fact that $\lim_{k \to \infty} \|x(k)\|_\infty = M^*$. Therefore, $\lim_{k \to \infty} |x_i(k)| = M^*$, for all $i \in V$.

Next, we show that $\lim_{k \to \infty} x_i(k)$ exists, for all $i \in V$. Without loss of generality, M^* is assumed to be nonzero and we fix any $i \in V$. Note that the fact that $\lim_{k \to \infty} |x_i(k)| = M^*$ include three possibilities: $\lim_{k \to \infty} x_i(k) = M^*$, $\lim_{k \to \infty} x_i(k) = -M^*$, or $x_i(k)$ switches between $-M^*$ and M^* infinitely as $k \to \infty$. The last possibility actually means $\liminf_{k \to \infty} x_i(k) = -M^*$ and $\limsup_{k \to \infty} x_i(k) = M^*$. We next prove the existence of the limit of $x_i(k)$ by showing that this last possibility cannot happen.

Suppose that we do have $\liminf_{k \to \infty} x_i(k) = -M^*$, and $\limsup_{k \to \infty} x_i(k) = M^*$. The following proof is based on a contradiction argument. We first use M^* and $x_i(k)$ to bound the trajectory of x_i after time instant k. Note that for all $s = 1, 2, \ldots$,

$$x_i(k+s) = \sum_{j \in \mathcal{N}_i(k+s-1) \cup \{i\}} a_{ij}(k+s-1)x_j(k+s-1)$$

$$\leq a_{ii}(k+s-1)x_i(k+s-1)$$

$$+ \sum_{j \in \mathcal{N}_i(k+s-1)} |a_{ij}(k+s-1)||x_j(k+s-1)|$$

$$\leq a_{ii}(k+s-1)x_i(k+s-1) + (1 - a_{ii}(k+s-1))\|x(k)\|_\infty.$$

It thus follows that $x_i(k+1) \leq \lambda x_i(k) + (1-\lambda)\|x(k)\|_\infty$. Therefore, for all $k \geq \widehat{k}(\varepsilon)$, it follows that $x_i(k+1) \leq \lambda x_i(k) + (1-\lambda)(M^* + \varepsilon)$. By a recursive analysis, we know that for all $k \geq \widehat{k}(\varepsilon)$ and all $s = 1, 2, \ldots$,

$$x_i(k+s) \leq \lambda^s x_i(k) + (1 - \lambda^s)(M^* + \varepsilon). \tag{7.7}$$

Since $\liminf_{k \to \infty} x_i(k) = -M^*$, for any given ε, there exists an infinite sequence $\{\bar{k}_\chi\}_{\chi=0}^\infty$ such that $\bar{k}_\chi > \widehat{k}(\varepsilon)$ and $x_i(\bar{k}_\chi) \leq -(M^* - \varepsilon)$, $\chi = 0, 1, \ldots$. In addition, since $\limsup_{k \to \infty} x_i(k) = M^*$, for any $\bar{k}_\chi \in \{\bar{k}_\ell\}_{\ell=0}^\infty$, there exists a time instant $\bar{k}_{\bar{\chi}} > \bar{k}_\chi$ such that $x_i(\bar{k}_{\bar{\chi}}) \geq M^* - \varepsilon$. By also noting that the state at each step is bounded by the previous step from (7.7), there must exist a time instant $\bar{k}_\chi^* \in$

$[\bar{k}_\chi, \bar{k}_{\bar{\chi}}]$ such that $-\lambda(M^* - \varepsilon) + (1 - \lambda)(M^* + \varepsilon) \leq x_i(\bar{k}_\chi^*) \leq -\lambda^2(M^* - \varepsilon) + (1 - \lambda^2)(M^* + \varepsilon)$.

Therefore, for all $\bar{k}_\chi$, it follows that

$$|x_i(\bar{k}_\chi^*)| \leq \max\{|(1 - 2\lambda)M^* + \varepsilon|, |(1 - 2\lambda^2)M^* + \varepsilon|\}$$

$$\leq M^* - (1 - \max\{|1 - 2\lambda|, |1 - 2\lambda^2|\})M^* + \varepsilon$$

$$< M^* - \frac{(1 - \max\{|1 - 2\lambda|, |1 - 2\lambda^2|\})M^*}{2},$$

if ε is chosen sufficiently small as

$$\varepsilon < \frac{(1 - \max\{|1 - 2\lambda|, |1 - 2\lambda^2|\})M^*}{2}.$$

This contradicts the fact that $\lim_{k \to \infty} |x_i(k)| = M^*$ and verifies that $\lim_{k \to \infty} x_i(k)$ exists.

Then, for undirected interaction graphs, we present the following result indicating that modulus consensus can be achieved with infinite joint connectivity.

Theorem 7.4 *Suppose that Assumption 7.1 holds and $\mathcal{G}_k$ is undirected for all $k \geq 0$. System (7.1) achieves asymptotic modulus consensus for all initial states $x(0) \in \mathbb{R}^N$ if $\{\mathcal{G}_k\}_0^\infty$ is infinitely jointly connected (see Definition 3.1).*

Proof Since $\mathcal{G}$ is infinitely jointly connected, the union graph $\mathcal{G}([k_1, \infty])$ is connected. We can therefore define $k_2 := \inf_k \{k \geq k_1, \mathcal{N}_{i_1}(k) \neq \emptyset\}$. We denote $\mathcal{V}_1 = \mathcal{N}_{i_1}(k_2)$. Obviously, we have that $|x_{i_1}(k_2)| = |x_{i_1}(k_1)| \leq M^* - \xi_1$, where $\xi_1 = M^* - \alpha_1 > 0$ is defined as in (7.3). Therefore, following a similar analysis by which we obtained (7.5) and (7.6), we know that $|x_i(k_2+1)| \leq M^* + \varepsilon - \lambda\xi_1$, $i \in \mathcal{V}_1$.

Similarly, since the union graph $\mathcal{G}([k_2 + 1, \infty])$ is connected, we can continue to define $k_3 := \inf_k \{k \geq k_2 + 1 : \bigcup_{i \in \mathcal{V}_1} (\mathcal{N}_i(k)) \neq \emptyset\}$. We also denote $\mathcal{V}_2 = \bigcup_{i \in \mathcal{V}_1} \mathcal{N}_i(k_3)$. Note that $\{i_1\} \subseteq \mathcal{V}_1 \subseteq \mathcal{V}_2$ with the definition of neighbor sets. The fact that the graph is undirected guarantees that k_3 is not only the first time instant that there is an edge from $\mathcal{V}_1$ to another node, but also the first time instant that there is an edge from another node to $\mathcal{V}_1$ during the time interval $[k_2 + 1, k_3]$. Therefore, we can apply Lemma 7.2 to the subset $\mathcal{V}_1$ for time interval $[k_2 + 1, k_3]$, and deduce that $|x_i(k_3)| \leq M^* + \varepsilon - \lambda\xi_1$, $i \in \mathcal{V}_1$. It then follows from the same analysis that $|x_i(k_3 + 1)| \leq M^* + \varepsilon - \lambda^2\xi_1$, $i \in \mathcal{V}_2$.

The above argument can be carried out recursively for $\mathcal{V}_3, \mathcal{V}_4, \ldots$ until $\mathcal{V}_m = \mathcal{V}$ for some constant $m \leq N - 1$. The corresponding k_m can be found based on infinite joint connectivity condition, where $|x_i(k_m + 1)| \leq M^* + \varepsilon - \lambda^m\xi_1$, for all $i \in \mathcal{V}$. This indicates that

$$\|x(k_m + 1)\|_\infty \leq M^* + \varepsilon - \lambda^m\xi_1 < M^* - \lambda^m\xi_1/2,$$

for sufficient small ε satisfying $\varepsilon < \lambda^{N-1}\xi_1/2$. This contradicts the fact that $\|x(k)\|_\infty \geq M^* - \varepsilon > M^* - \lambda^m\xi_1/2, \forall k \geq \widehat{k}(\varepsilon)$. Therefore, it follows that $\lim_{k\to\infty} |x_i(k)| = M^*$, for all $i \in \mathcal{V}$.

Also, using the same analysis in Theorem 7.3, we can show that $\lim_{k\to\infty} x_i(k)$ exists, for all $i \in \mathcal{V}$. To this end, we have proven Theorem 7.4.

7.4 Simulations

Consider the following discrete-time Kuramoto oscillator systems with antagonistic and cooperative edges:

$$\theta_i(k+1) = \theta_i(k) - \mu \sum_{j\in\mathcal{N}_i(k)} \sin\Big(\theta_i(k) - R_{ij}(k)\theta_j(k)\Big), \qquad (7.8)$$

where $\theta_i(k)$ denotes the state of node i at time k, $\mu > 0$ is the stepsize, and $R_{ij}(k) \in \{1, -1\}$ represents the cooperative or antagonistic relationship between node i and node j. Note that with $R_{ij}(k) \equiv 1$, system (7.8) corresponds to the classical Kuramoto oscillator model [7, 21]. Let $\sigma \in (0, \frac{\pi}{2})$ be a given constant and suppose $\theta_i(0) \in (-\frac{\pi}{2}+\sigma, \frac{\pi}{2}-\sigma)$ for all $i \in \mathcal{V}$. Here σ can be any positive constant sufficiently small.

System (7.8) can be rewritten as

$$\theta_i(k+1) = \theta_i(k) - \mu \sum_{j\in\mathcal{N}_i(k)} \frac{\sin(\theta_i(k) - R_{ij}(k)\theta_j(k))}{\theta_i(k) - R_{ij}(k)\theta_j(k)}(\theta_i(k) - R_{ij}(k)\theta_j(k)).$$

Note that the function $\frac{\sin x}{x}$ is well-defined for $x \in (-\infty, \infty)$. Therefore, defining

$$a_{ij}(\theta, k) = \frac{\sin(\theta_i(k) - R_{ij}(k)\theta_j(k))}{\theta_i(k) - R_{ij}(k)\theta_j(k)}R_{ij}(k), \quad j \in \mathcal{N}_i(k),$$

and $a_{ii}(\theta, k) = 1 - \mu \sum_{j\in\mathcal{N}_i(k)} |a_{ij}(\theta, k)|$, where $\theta = [\theta_1, \theta_2, \ldots, \theta_N]^\mathrm{T}$, system (7.8) is rewritten into the form of (7.1).

Moreover, Lemma 7.2 ensures that

$$0 < \lambda^* \leq \frac{\sin(\theta_i(k) - R_{ij}(k)\theta_j(k))}{\theta_i(k) - R_{ij}(k)\theta_j(k)} \leq 1,$$

where $\lambda^* = \frac{\sin(\pi - 2\sigma)}{\pi - 2\sigma}$. This gives us $|a_{ij}(\theta, k)| \geq \lambda^*$, for all i and $j \in \mathcal{N}_i(k)$. In addition, by selecting $\mu < \frac{1-\lambda^*}{N}$, we can also guarantee that $|a_{ii}(\theta, k)| \geq \lambda^*$ for all k.

Fig. 7.1 Switching
communication topology

$$\mathcal{G}_1 \longrightarrow \mathcal{G}_2 \longrightarrow \mathcal{G}_3 \longrightarrow \mathcal{G}_1 \longrightarrow \cdots$$

$$1 \xleftarrow{\quad - \quad} 3$$

2

Fig. 7.2 $\mathcal{G}_1$

1   3

Fig. 7.3 $\mathcal{G}_2$

1   3

Fig. 7.4 $\mathcal{G}_3$

Therefore, given $\mu < \frac{1-\lambda^*}{N}$, modulus consensus, *i.e.*, $\lim_{k\to\infty}(|\theta_i(k)| - |\theta_j(k)|) = 0$, $i, j \in \mathcal{V}$, is achieved for system (7.8), if $\{\mathcal{G}_k\}_0^\infty$ is uniformly jointly strongly connected for directed graphs, or infinitely jointly connected for undirected graphs according to Theorems 7.3 and 7.4.

We next verify the above arguments using simulations. For the case of directed communication topology, we assume that the communication topology switches periodically as Fig. 7.1 when the systems are at time instants $\eta_\ell = \ell s$, $\ell = 1, 2, \ldots,$ where $\mathcal{G}_1, \mathcal{G}_2, \mathcal{G}_3$ are represented in Figs. 7.2, 7.3, and 7.4.

The signed weight matrices associated with $\mathcal{G}_1, \mathcal{G}_2, \mathcal{G}_3$ are given by

$$A_1 = \begin{bmatrix} 1 & 0 & 0 \\ 0 & 1 & 0 \\ -0.5 & 0 & 0.5 \end{bmatrix}, \quad A_2 = \begin{bmatrix} 1 & 0 & 0 \\ 0 & 0.5 & -0.5 \\ 0 & 0 & 1 \end{bmatrix}, \quad A_3 = \begin{bmatrix} 0.5 & 0.5 & 0 \\ 0 & 1 & 0 \\ 0 & 0.5 & 0.5 \end{bmatrix}.$$

The initial states are $x_1(0) = -1.5$, $x_2(0) = 1$, and $x_3(0) = 0$, and μ is chosen as $\mu = 0.1$. Figure 7.5 shows the state evolutions of states over directed switching

Fig. 7.5 Modulus consensus for the directed communication topology

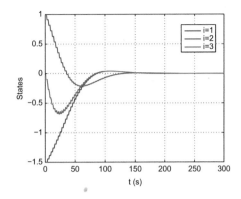

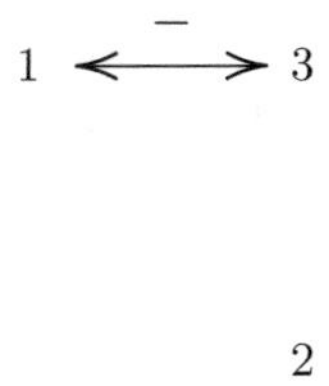

1 $\longleftrightarrow$ 3

2

Fig. 7.6 $\mathcal{G}_4$

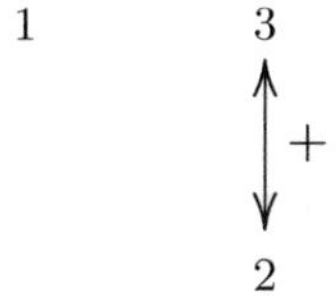

1   3

$+$

2

Fig. 7.7 $\mathcal{G}_5$

communication topologies. We see that modulus consensus is achieved for this group of oscillators with antagonistic interactions and switching topologies, in accordance with the conclusion from Theorem 7.3. Note that states of all the agents converge to zero, instead of achieving bipartite consensus.

For the case of undirected communication topology, we assume the communication topology switches between $\mathcal{G}_4$ and $\mathcal{G}_5$ in Figs. 7.6 and 7.7. The topology remains $\mathcal{G}_4$ but at time intervals $[\ell^2, \ell^2 + 1]$, where the topology is $\mathcal{G}_5$, $\ell = 1, 2, \ldots$. The signed weight matrices associated with $\mathcal{G}_4, \mathcal{G}_5$ are

$$A_4 = \begin{bmatrix} 0.5 & 0 & -0.5 \\ 0 & 1 & 0 \\ -0.5 & 0 & 0.5 \end{bmatrix}, \quad A_5 = \begin{bmatrix} 1 & 0 & 0 \\ 0 & 0.5 & 0.5 \\ 0 & 0.5 & 0.5 \end{bmatrix}.$$

The initial states are $x_1(0) = -1.5$, $x_2(0) = 1$, and $x_3(0) = 0$, and μ is chosen as $\mu = 0.1$. Figure 7.8 shows the state evolutions over undirected switching

Fig. 7.8 Modulus consensus
for the undirected
communication topology

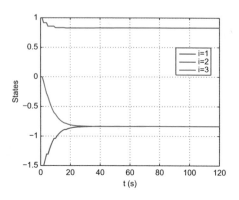

communication topologies. We see that modulus consensus is achieved for this group of oscillators with antagonistic interactions and switching topologies, in accordance with the conclusion from Theorem 7.4. More specifically, bipartite consensus is achieved in this case.

7.5   Literature

Distributed consensus algorithms are extensively studied in recent years [4, 11, 17, 18]. In particular, both continuous-time and discrete-time models are studied for consensus algorithms with switching interaction graphs and many in-depth understanding is obtained for linear models [5, 12, 16]. Nonlinear networked dynamical systems have also drawn much attention [13, 14, 19] since in many practical problems the node dynamics are naturally nonlinear, see, e.g., the Kuramoto model [7, 21].

Motivated by opinion dynamics over social networks [6, 8, 9], consensus algorithms over cooperative-antagonistic networks have drawn much attention [1, 2, 20]. The results in this chapter are based mainly on [15]. For further results on modulus and bipartite consensus of coupled dynamical systems, see [2, 3, 10, 20, 22]. In particular, Altafini [2] assumes that a node receives the opposite of the true state of its neighboring node if they are antagonistic. Therefore, the modeling of such an antagonistic input for agent i is of the form $-(x_i + x_j)$ (in contrast to the form $-(x_i - x_j)$ for cooperative input), where j denotes the antagonistic neighbor of agent i. On the other hand, the authors of [20] assume that a node receives the opposite of the relative state from its neighboring node if they are antagonistic. Then, the antagonistic input for agent i is modeled of the form $(x_i - x_j)$ in this case. The extension to the case of homogenous single-input high-order dynamical systems is discussed in [22]. The graph is assumed to be fixed and a spectral analysis approach is used. In addition, a lifting approach is proposed in [10] to study opinion dynamics with antagonisms over switching graphs. Some general conditions are established by applying the rich results from the consensus literature. The dissensus problem is studied in [3], where the focus is to understand when consensus is or is not achieved if death and duplication phenomena occur. Note that in dissensus the control terms for death and duplication phenomena are added to the classical consensus algorithm.

Acknowledgments ©2016 Elsevier. Reprinted, with permission, from Ziyang Meng, Guodong Shi, K. H. Johansson, Ming Cao, Yiguang Hong, "Behaviors of networks with antagonistic interactions and switching topologies", Automatica, vol. 73, pp. 110–116, 2016.

References

1. C. Altafini, Dynamics of opinion forming in structurally balanced social networks. PLoS One **7**(6), 1–8 (2012)
2. C. Altafini, Consensus problems on networks with antagonistic interactions. IEEE Trans. Autom. Control **58**(4), 935–946 (2013)
3. D. Bauso, L. Giarre, R. Pesenti, Quantized dissensus in networks of agents subject to death and duplication. IEEE Trans. Autom. Control **57**(3), 783–788 (2012)
4. V.D. Blondel, J.M. Hendrickx, A. Olshevsky et al., Convergence in multiagent coordination, consensus, and flocking, in *Proceedings of the 44th IEEE Conference on Decision and Control*, Seville, 2005, pp. 2996–3000
5. M. Cao, A.S. Morse, B.D.O. Anderson, Reaching a consensus in a dynamically changing environment: a graphical approach. SIAM J. Control Optim. **47**(2), 575–600 (2008)
6. D. Cartwright, F. Harary, Structural balance: a generalization of Heiders theory. Psychol. Rev. **63**(5), 277–293 (1956)
7. N. Chopra, M.W. Spong, On exponential synchronization of Kuramoto oscillators. IEEE Trans. Autom. Control **54**(2), 353–357 (2009)
8. D. Easley, J. Kleinberg, *Networks, Crowds, and Markets: Reasoning About a Highly Connected World* (Cambridge University Press, Cambridge, 2010)
9. R. Hegselmann, U. Krause, Opinion dynamics and bounded confidence models, analysis, and simulation. J. Artif. Soc. Soc. Simul. **5**(3), 1–33 (2002)
10. J.M. Hendrickx, A lifting approach to models of opinion dynamics with antagonisms, in *Proceedings of the IEEE Conference on Decision and Control*, Los Angeles, 2014, pp. 2118–2123
11. J.M. Hendrickx, J.N. Tsitsiklis, Convergence of type-symmetric and cut-balanced consensus seeking systems. IEEE Trans. Autom. Control **58**(1), 214–218 (2013)
12. A. Jadbabaie, J. Lin, A.S. Morse, Coordination of groups of mobile autonomous agents using nearest neighbor rules. IEEE Trans. Autom. Control **48**(6), 988–1001 (2003)
13. Z. Lin, B. Francis, M. Maggiore, State agreement for continuous-time coupled nonlinear systems. SIAM J. Control Optim. **46**(1), 288–307 (2007)
14. Z. Meng, Z. Lin, W. Ren, Robust cooperative tracking for multiple non-identical second-order nonlinear systems. Automatica **49**(8), 2363–2372 (2013)
15. Z. Meng, G. Shi, K.H. Johansson et al., Behaviors of networks with antagonistic interactions and switching topologies. Automatica **73**, 110–116 (2016)
16. L. Moreau, Stability of continuous-time distributed consensus algorithms, in *Proceedings of the IEEE Conference on Decision and Control*, Nassau, Bahamas, 2004, pp. 3998–4003
17. R. Olfati-Saber, R.M. Murray, Consensus problems in networks of agents with switching topology and time-delays. IEEE Trans. Autom. Control **49**(9), 1520–1533 (2004)
18. W. Ren, R.W. Beard, Consensus seeking in multiagent systems under dynamically changing interaction topologies. IEEE Trans. Autom. Control **50**(5), 655–661 (2005)
19. G. Shi, Y. Hong, Global target aggregation and state agreement of nonlinear multi-agent systems with switching topologies. Automatica **45**(5), 1165–1175 (2009)

20. G. Shi, M. Johansson, K.H. Johansson, How agreement and disagreement evolve over random dynamic networks. IEEE J. Select. Areas Commun. **31**(6), 1061–1071 (2013)
21. S.H. Strogatz, From Kuramoto to Crawford: exploring the onset of synchronization in populations of coupled oscillators. Phys. D: Nonlinear Phenom. **143**(1–4), 1–20 (2000)
22. M.E. Valcher, P. Misra, On the consensus and bipartite consensus in high-order multi-agent dynamical systems with antagonistic interactions. Syst. Control Lett. **66**(3), 94–103 (2014)
23. S. Wasserman, K. Faust, *Social Network Analysis: Methods and Applications* (Cambridge University Press, Cambridge, 1994)

Part III
Control

Chapter 8
Control of Networked Dynamical System with Input Saturation

In almost every physical application, the actuator has bounds on its input, and thus actuator saturation is important to study. Also, it is quite possible that the system becomes unstable when input saturation constraints are considered. In this chapter, we consider the global consensus problem for discrete-time networked dynamical systems with input saturation constraints under fixed undirected graphs. We first give necessary conditions for achieving the global consensus via a distributed protocol based on relative state measurements of the agent itself and its neighboring agents. We then focus on two special cases, where the agent dynamics are either neutrally stable or double integrator. For the neutrally stable case, any linear protocol of a particular form, which solves the consensus problem for the case without input saturation constraints, also solves the global consensus problem for the case with input saturation constraints. For the double integrator case, we show that a subset of linear protocols, which solve the consensus problem for the case without input saturation constraints, also solve the global consensus problem for the case with input saturation constraints. The results are illustrated by numerical simulations.

8.1 Problem Formulation and Control Algorithm

In almost every physical application, the actuator has bounds on its input. We specify (3.2) into the following equation to study the influence of the constraints on inputs.

$$x_i(k+1) = \tilde{A}x_i(k) + \tilde{B}\sigma(u_i(k)), \quad i \in \mathcal{V}, \tag{8.1}$$

where $x_i(k) \in \mathbb{R}^n$, $u_i(k) \in \mathbb{R}^m$, $\tilde{A}$ and $\tilde{B}$ are compatible matrices,

$$\sigma(u_i(k)) = [\sigma_1(u_{i,1}(k)), \sigma_1(u_{i,2}(k)), \ldots, \sigma_1(u_{i,m}(k))]^{\mathrm{T}},$$

© The Author(s), under exclusive license to Springer Nature Switzerland AG 2021
Z. Meng et al., *Modelling, Analysis, and Control of Networked Dynamical Systems*,
Systems & Control: Foundations & Applications,
https://doi.org/10.1007/978-3-030-84682-4_8

and each $\sigma_1(u)$ is the standard saturation function, i.e.,

$$\sigma_1(u) = \begin{cases} 1 & \text{if } u > 1, \\ u & \text{if } |u| \leq 1, \\ -1 & \text{if } u < -1. \end{cases}$$

The only information available for agent i comes from the network. In particular, agent i receives a linear combination of its own states relative to that of neighboring agents, i.e.,

$$\zeta_i(k) = \sum_{j \in \mathcal{N}_i} a_{ij}(x_i(k) - x_j(k)).$$

Our goal is to design distributed protocols $u_i(k)$ for $i \in \mathcal{V}$ by using $\zeta_i(k)$ to solve the global consensus problem, i.e., for any initial condition $x_i(0)$, $i \in \mathcal{V}$, $\lim_{k \to \infty}(x_i(k) - x_j(k)) = 0$ for all $i, j \in \mathcal{V}$.

Note that each agent is subject to the input saturation constraints. These nonlinearities make the protocol design for achieving consensus difficult since we have to guarantee that consensus is achieved for all initial conditions. In order to achieve consensus for (8.1), we first present some necessary conditions.

Assumption 8.1 *The agent model* (8.1) *is asymptotically null controllable with bounded controls (ANCBC), i.e., the pair* $(\tilde{A}, \tilde{B})$ *is stabilizable and all the eigenvalues of the matrix* $\tilde{A}$ *are within or on the unit circle.*

From the saturation literature [8, 9], it is evident that, in general, we need to design a nonlinear protocol to solve the global consensus problem. In this chapter, we shall concentrate on a linear protocol

$$u_i(k) = K\zeta_i(k) = K \sum_{j \in \mathcal{N}_i} a_{ij}(x_i(k) - x_j(k)), \quad i \in \mathcal{V}, \tag{8.2}$$

as such a protocol may suffice in some cases.

Given a fixed undirected graph and Assumption 8.1, it follows from [15, Theorem 3.1] that a network of N agents (8.1) in the absence of input saturation achieves consensus via the protocol (8.2) if and only if the following assumption is satisfied.

Assumption 8.2 *The graph* $\mathcal{G}$ *is connected.*

Based on the results in [11] and [5, Section 4.2], we have the following result.

Theorem 8.1 *A networked dynamical system of N agents* (8.1) *achieves the global consensus via distributed protocols* (8.2) *only if Assumptions 8.1 and 8.2 are satisfied.*

It is known that for some special discrete-time cases, such as neutrally stable systems [1], and a double integrator [12], there exist saturated globally stabilizing linear state feedback control laws. Hence, in the following sections, we consider the global consensus problem for such special cases. We show that Assumptions 8.1 and 8.2 are also sufficient for achieving the global consensus for these cases by explicitly specifying the matrix K for (8.2).

8.2 Neutrally Stable Agent Dynamics

In this section, we consider the case where the agent dynamics (8.1) is open-loop neutrally stable.

Under Assumption 8.1, there exists a non-singular state transformation T^{-1}, such that

$$\tilde{A} = T^{-1} \begin{bmatrix} \tilde{A}_c & 0 \\ 0 & \tilde{A}_s \end{bmatrix} T, \quad \tilde{B} = T^{-1} \begin{bmatrix} \tilde{B}_c \\ \tilde{B}_s \end{bmatrix},$$

where $\tilde{A}_c^T \tilde{A}_c = I$, $\tilde{A}_s$ is Schur stable (i.e., all its eigenvalues are within the unit circle), and the pair $(\tilde{A}_c, \tilde{B}_c)$ is controllable.

As shown in [15], the asymptotically stable modes can be ignored since we can set the corresponding gain matrix to zero. Thus, without loss of generality, we make the following assumption in this section.

Assumption 8.3 $\tilde{A}^T \tilde{A} = I_n$ and the pair $(\tilde{A}, \tilde{B})$ is controllable.

Under Assumption 8.3, controllability of the pair $(\tilde{A}, \tilde{B})$ is equivalent to stabilizability of the pair $(\tilde{A}, \tilde{B})$.

Consider the following control law:

$$u_i(k) = -\varepsilon \tilde{B}^T \tilde{A} \sum_{j \in \mathcal{N}_i} a_{ij}(x_i(k) - x_j(k)), \quad i \in \mathcal{V}. \tag{8.3}$$

Note that (8.3) is of the form (8.2) with $K = -\varepsilon \tilde{B}^T \tilde{A}$, where ε is a designed parameter. The following lemma shows that the protocol (8.3) with a properly chosen ε solves the consensus problem for a networked dynamical system without input saturation.

Lemma 8.2 *Consider a networked dynamical system of N agents (8.1) in the absence of input saturation constraints. Assume that Assumptions 8.2 and 8.3 are satisfied. Then any protocol (8.3) with $\varepsilon \in (0, \frac{2}{\lambda_N \|\tilde{B}^T \tilde{B}\|})$, where λ_N is the largest eigenvalue of the corresponding Laplacian matrix, solves the consensus problem.*

Proof It is well known from [6, 16] that the consensus problem for a network of N identical agents is equivalent to the simultaneous stabilization problem of $N - 1$

systems. Hence, it can be verified that consensus is achieved via (8.3) if all the matrices $\tilde{A} - \varepsilon \lambda_i \tilde{B} \tilde{B}^T \tilde{A}$, where λ_i, $i \in \{2, \ldots, N\}$ (i.e., the nonzero eigenvalues of the Laplacian matrix) are Schur stable. It then follows from [7, Lemma 4.2] that all these matrices are Schur stable if $\varepsilon \in (0, \frac{2}{\lambda_N \|\tilde{B}^T \tilde{B}\|})$.

The following theorem shows that (8.3) with $\varepsilon \in (0, \frac{2}{\lambda_N \|\tilde{B}^T \tilde{B}\|})$ also solves the global consensus problem for the networked dynamical system in the presence of input saturation constraints.

Theorem 8.3 *Consider a networked dynamical system of N agents (8.1). Assume that Assumptions 8.2 and 8.3 are satisfied. Then any protocol (8.3) with $\varepsilon \in (0, \frac{2}{\lambda_N \|\tilde{B}^T \tilde{B}\|})$ solves the global consensus problem.*

Proof Define $x(k) = [x_1^T(k), \ldots, x_N^T(k)]^T$ and $u(k) = [u_1^T(k), \ldots, u_N^T(k)]^T$. To simplify the notation, sometimes x or u without explicitly indicating the time instant will refer to $x(k)$ or $u(k)$, respectively. With these definitions, we obtain the following dynamics:

$$x(k+1) = (I_N \otimes \tilde{A})x(k) + (I_N \otimes \tilde{B})\sigma(u(k)), \tag{8.4a}$$

$$u(k) = -\varepsilon(L \otimes \tilde{B}^T \tilde{A})x(k). \tag{8.4b}$$

Motivated by the Lyapunov function in [2, 17], we consider the Lyapunov candidate

$$V(x(k)) = \frac{1}{2} x^T(k)(L \otimes I_n)x(k).$$

Define a manifold where all the agent states are identical

$$\mathcal{M} := \{x \in \mathbb{R}^{Nn} | x_1 = x_2 = \ldots = x_N\}.$$

Note that $V(x) \geq 0$ and $V(x) = 0$ if and only if $x \in \mathcal{M}$. Let us now evaluate $\Delta V(x(k)) = V(x(k+1)) - V(x(k))$. We sometimes drop the dependency of $V(x(k))$ and $\Delta V(x(k))$ on $x(k)$ for notational simplification when it is clear from the context. From the dynamics of (8.4), we obtain

$$\Delta V = \frac{1}{2}\sigma^T(u)(L \otimes \tilde{B}^T \tilde{A})x + \frac{1}{2}x^T(L \otimes \tilde{A}^T \tilde{B})\sigma(u) + \frac{1}{2}\sigma^T(u)(L \otimes \tilde{B}^T \tilde{B})\sigma(u)$$

$$= -\frac{1}{\varepsilon}\sigma^T(u)u + \frac{1}{2}\sigma^T(u)(L \otimes \tilde{B}^T \tilde{B})\sigma(u)$$

$$\leq -\sigma^T(u)(\frac{1}{\varepsilon}I_{Nm} - \frac{1}{2}L \otimes \tilde{B}^T \tilde{B})\sigma(u),$$

where we have used that $L = L^T$ for undirected graphs and that $z^T\sigma(z) \geq \sigma^T(z)\sigma(z)$ for any column vector z.

Since $\varepsilon \in (0, \frac{2}{\lambda_N \|\tilde{B}^\mathrm{T}\tilde{B}\|})$, $\Delta V \leq 0$ and $\Delta V = 0$ if and only if $(L \otimes \tilde{B}^\mathrm{T}\tilde{A})x = 0$. We shall show that $(L \otimes \tilde{B}^\mathrm{T}\tilde{A})x = 0$ if and only if $x \in \mathcal{M}$, which in turn implies that $\Delta V = 0$ if and only if $x \in \mathcal{M}$. We first note that if $x \in \mathcal{M}$, then $(L \otimes \tilde{B}^\mathrm{T}\tilde{A})x = 0$ since the graph is connected. We then show that $(L \otimes \tilde{B}^\mathrm{T}\tilde{A})x = 0$ implies that $x \in \mathcal{M}$. Note that $(L \otimes \tilde{B}^\mathrm{T}\tilde{A})x = 0$ implies that $(\tilde{L} \otimes \tilde{B}^\mathrm{T}\tilde{A})q = 0$, where the relative state $q = [q_2, \ldots, q_N]^\mathrm{T}$, $q_i = x_i - x_1$ for $i \in \{2, \ldots, N\}$, and

$$\tilde{L} = \begin{bmatrix} l_{2,2} - l_{1,2} & \cdots & l_{2,N} - l_{1,N} \\ \vdots & \ddots & \vdots \\ l_{N,2} - l_{1,2} & \cdots & l_{N,N} - l_{1,N} \end{bmatrix} \in \mathbb{R}^{(N-1)\times(N-1)}. \qquad (8.5)$$

Since the graph is connected, it follows from [16, Lemma 1] that the eigenvalues of $\tilde{L}$ are the nonzero eigenvalues of the matrix L, which are positive. Thus, the matrix $\tilde{L}$ is non-singular, i.e., $\mathrm{rank}(\tilde{L}) = N - 1$.

From the fact that $\tilde{A}^\mathrm{T}\tilde{A} = I_n$, we see that $(\tilde{L} \otimes \tilde{B}^\mathrm{T}\tilde{A})q = 0$ implies $q^\mathrm{T}(\tilde{L} \otimes \tilde{A}^{-1}\tilde{B}) = 0$. Also note that $q(k+1) = (I_{N-1} \otimes \tilde{A})q(k)$, since $u(k) = -\varepsilon(L \otimes \tilde{B}^\mathrm{T}\tilde{A})x(k) = 0$. Therefore

$$(\tilde{L} \otimes \tilde{B}^\mathrm{T}\tilde{A})q(k+1) = (\tilde{L} \otimes \tilde{B}^\mathrm{T}\tilde{A}^2)q(k),$$

which is equivalent to $q^\mathrm{T}(\tilde{L} \otimes \tilde{A}^{-2}\tilde{B}) = 0$. By iteration, we obtain $q^\mathrm{T}(\tilde{L} \otimes \tilde{A}^{-r}\tilde{B}) = 0$ for $r = 3, 4, \ldots, n+1$. Hence,

$$q^\mathrm{T}\left(\tilde{L} \otimes \tilde{A}^{-(n+1)}\left[\tilde{A}^n\tilde{B} \ldots \tilde{A}\tilde{B} \ \tilde{B}\right]\right) = 0. \qquad (8.6)$$

Note that $\mathrm{rank}\left(\left[\left[\tilde{A}^n\tilde{B} \ldots \tilde{A}\tilde{B} \ \tilde{B}\right]\right]\right) = n$ since the pair $(\tilde{A}, \tilde{B})$ is controllable. This together with the fact that the matrix $\tilde{A}$ is non-singular implies that $\mathrm{rank}\left(\tilde{A}^{-(n+1)}\left[\left[\tilde{A}^n\tilde{B} \ldots \tilde{A}\tilde{B} \ \tilde{B}\right]\right]\right) = n$. Finally, using the property of Kronecker product, we obtain

$$\mathrm{rank}\left(\tilde{L} \otimes \tilde{A}^{-(n+1)}\left[\tilde{A}^n\tilde{B} \ldots \tilde{A}\tilde{B} \ \tilde{B}\right]\right)$$

$$= \mathrm{rank}\left(\tilde{L}\right)\mathrm{rank}\left(\tilde{A}^{-(n+1)}\left[\tilde{A}^n\tilde{B} \ldots \tilde{A}\tilde{B} \ \tilde{B}\right]\right) = (N-1)n.$$

Therefore, the only solution of (8.6) is $q = 0$, which is equivalent to $x_1 = \ldots = x_N$, i.e., $x \in \mathcal{M}$. Hence, we have shown that $\Delta V(x) \leq 0$ and $\Delta V(x) = 0$ if and only if $x \in \mathcal{M}$.

Since $\Delta V(x(k)) = V(x(k+1)) - V(x(k)) \leq 0$, we conclude that $V(x(k))$ is non-increasing in k. Thus, $\lim_{k\to\infty} V(x(k)) = V_\star$ for some $V_\star \geq 0$ since $V \geq 0$. This implies that $\Delta V(x(k)) \to 0$ as $k \to \infty$ and hence $x(k) \to \mathcal{M}$ as $k \to \infty$. Hence, the global consensus is achieved.

8.3 Double Integrator Agent Dynamics

In this section, we consider the case where the agent model (8.1) is a double integrator.

Assumption 8.4 *The matrices $\tilde{A}$ and $\tilde{B}$ are of the form*

$$\tilde{A} = \begin{bmatrix} 1 & 1 \\ 0 & 1 \end{bmatrix}, \quad \tilde{B} = \begin{bmatrix} 0 \\ 1 \end{bmatrix}.$$

Let us first recall the following result that gives a necessary and sufficient condition on the feedback gain parameters for achieving consensus without input saturation constraints.

Lemma 8.4 ([10]) *Consider a networked dynamical system of N agents described by*

$$\begin{bmatrix} x_i(k+1) \\ v_i(k+1) \end{bmatrix} = \tilde{A} \begin{bmatrix} x_i(k) \\ v_i(k) \end{bmatrix} + \tilde{B} u_i(k), \quad i \in \mathcal{V}. \tag{8.7}$$

Assume that Assumptions 8.2 and 8.4 are satisfied. Then the protocol (8.2) with $K = -\begin{bmatrix} [\alpha \ \beta] \end{bmatrix}$:

$$u_i(k) = -\alpha \sum_{j \in \mathcal{N}_i} a_{ij}(x_i(k) - x_j(k)) - \beta \sum_{j \in \mathcal{N}_i} a_{ij}(v_i(k) - v_j(k)), \tag{8.8}$$

solves the consensus problem if and only if

$$0 < \alpha < \beta < \frac{\alpha}{2} + \frac{2}{\lambda_N}. \tag{8.9}$$

The following theorem shows that a subset of the protocols (8.8), which solve the consensus problem for a networked dynamical system without input saturation constraints, also solve the global consensus problem for a networked dynamical system with input saturation constraints.

Theorem 8.5 *Consider a networked dynamical system of N agents (8.1). Assume that Assumptions 8.2 and 8.4 are satisfied. Then the protocol (8.8) with*

$$0 < \sqrt{3}\alpha < \beta < \frac{3}{2\lambda_N} \tag{8.10}$$

solves the global consensus problem.

Proof Let $x(k) = [x_1^T(k), \ldots, x_N^T(k)]^T$, $v(k) = [v_1^T(k), \ldots, v_N^T(k)]^T$, $u(k) = [u_1^T(k), \ldots, u_N^T(k)]^T$, $y_i(k) = [x_i^T(k), v_i^T(k)]^T$, and $y(k) = [y_1^T(k), \ldots, y_N^T(k)]^T$.

To simplify the notation, sometimes x, v, u, or y without explicitly indicating the time instant will refer to $x(k)$, $v(k)$, $u(k)$, or $y(k)$, respectively. With these definitions, we obtain the following dynamics:

$$y(k+1) = (I_N \otimes \tilde{A})y(k) + (I_N \otimes \tilde{B})\sigma(u(k)),$$

$$u(k) = \left(L \otimes [[-\alpha \ -\beta]]\right) y(k).$$

We also obtain that $x(k+1) = x(k) + v(k)$ and $v(k+1) = v(k) + \sigma(u(k))$. Note that $u(k)$ can be written in terms of $x(k)$ and $v(k)$ as

$$u(k) = -\alpha L x(k) - \beta L v(k). \tag{8.11}$$

Hence, we obtain

$$u(k+1) = u(k) - \alpha L v(k) - \beta L \sigma(u(k)). \tag{8.12}$$

Motivated by the Lyapunov function in [12], we consider the following Lyapunov candidate:

$$V(y) = -\sigma^{\mathrm{T}}(u)\sigma(u) + 2\sigma^{\mathrm{T}}(u)u + 2\alpha\sigma^{\mathrm{T}}(u)Lv + \alpha v^{\mathrm{T}}Lv.$$

We sometimes drop the dependency of $V(y(k))$ on $y(k)$ for notation simplification when it is clear from the context. Similar to the proof of Theorem 8.3, we define a manifold where all the agent states are identical

$$\mathcal{M} := \{y \in \mathbb{R}^{2N} | y_1 = y_2 = \ldots = y_N\}.$$

Note that $V(y) = 0$ if $y \in \mathcal{M}$. We will show that $V(y) \geq 0$ and $V(y) = 0$ if only if $y \in \mathcal{M}$. Since $\sigma^{\mathrm{T}}(z)z \geq \sigma^{\mathrm{T}}(z)\sigma(z)$ and the equality holds if and only if $-\mathbf{1_m} \leq z \leq \mathbf{1_m}$, we obtain

$$V \geq \sigma^{\mathrm{T}}(u)\sigma(u) + 2\alpha\sigma^{\mathrm{T}}(u)Lv + \alpha v^{\mathrm{T}}Lv \tag{8.13}$$

$$= \left[\begin{bmatrix} \sigma(u) \\ Lv \end{bmatrix}\right]^{\mathrm{T}} \begin{bmatrix} 1 & \alpha \\ \alpha & \frac{2}{3}\alpha\beta \end{bmatrix} \left[\begin{bmatrix} \sigma(u) \\ Lv \end{bmatrix}\right] + v^{\mathrm{T}}(\alpha L - \frac{2}{3}\alpha\beta L^{\mathrm{T}}L)v, \tag{8.14}$$

where the equality of (8.13) holds if and only if $-\mathbf{1_m} \leq u \leq \mathbf{1_m}$. Since $\beta > \sqrt{3}\alpha > \frac{3}{2}\alpha > 0$, the first term of (8.14) is non-negative and is equal to zero if and only if $\sigma(u) = 0$ and $Lv = 0$. Note that from (8.11), we see that $u = -\alpha Lx - \beta Lv = -\alpha Lx$; therefore, $Lx = 0$ since $\alpha \neq 0$. Thus, the first term is equal to zero if and only if $y \in \mathcal{M}$. We next show that the second term is also non-negative. Since $L = L^{\mathrm{T}}$, we see that the eigenvalues of the matrix $\alpha L - \frac{2}{3}\alpha\beta L^{\mathrm{T}}L$ are $\alpha\lambda_i(1 - \frac{2}{3}\beta\lambda_i)$, where λ_i, $i \in \{1, \ldots, N\}$, are the eigenvalues of the Laplacian matrix L. Since

$\beta \lambda_N < \frac{3}{2}$, the second term is non-negative and equal to zero if and only if $Lv = 0$. Therefore, $V(y) \geq 0$ and $V(y) = 0$ if and only if $y \in \mathcal{M}$.

Next, we show that $\Delta V(y(k)) = V(y(k+1)) - V(y(k)) \leq 0$. We sometimes drop the dependency $\Delta V(y(k))$ on $y(k)$ for notational simplification when it is clear from the context. With some algebra, we obtain

$$V(y(k+1)) = -t^T t + 2t^T u + 2(\alpha - \beta)t^T L\sigma(u) + \alpha v^T Lv$$
$$+ 2\alpha v^T L\sigma(u) + \alpha \sigma^T(u)L\sigma(u),$$

where to simplify notation we have used $t = \sigma(u(k+1))$. Note that $-\mathbf{1_m} \leq t \leq \mathbf{1_m}$ by the definition of the saturation function. Thus,

$$\Delta V(y(k))$$
$$= V(y(k+1)) - V(y(k))$$
$$= -t^T t + 2t^T u + 2(\alpha - \beta)t^T L\sigma(u) + \sigma^T(u)(\alpha L + I)\sigma(u) - 2\sigma^T(u)u.$$

Without loss of generality, we assume that $u_i > 1$ for $i \in \{1, \ldots, N_1\} := S_p$, $|u_i| \leq 1$ for $i \in \{N_1 + 1, \ldots, N_2\} := S_m$, and $u_i < -1$ for $i \in \{N_2 + 1, \ldots, N\} := S_q$, since if this is not the case, we can always relabel the nodes to achieve this. Note that the sets S_p, S_m, and S_q may be empty. We then define the partition $t = [t_p; t_m; t_q]$, $u = [u_p; u_m; u_q]$, where $t_p, u_p \in \mathbb{R}^{N_1}$, $t_m, u_m \in \mathbb{R}^{N_2 - N_1}$, and $t_q, u_q \in \mathbb{R}^{N - N_2}$ are defined accordingly. We partition the Laplacian matrix L accordingly

$$L = \begin{bmatrix} L_{pp} & L_{pm} & L_{pq} \\ L_{pm}^T & L_{mm} & L_{mq} \\ L_{pq}^T & L_{mq}^T & L_{qq} \end{bmatrix},$$

where L_{pp}, L_{pm}, L_{pq}, L_{mm}, L_{mq}, and L_{qq} are real matrices of appropriate dimensions. With some algebra, we obtain

$$\Delta V = 2(t_p - \mathbf{1}_p)^T(u_p - \mathbf{1}_p) + 2t_p^T \mathbf{1}_p - 2\mathbf{1}_p^T \mathbf{1}_p$$
$$+ 2(t_q + \mathbf{1}_q)^T(u_q + \mathbf{1}_q) - 2t_q^T \mathbf{1}_q - 2\mathbf{1}_q^T \mathbf{1}_q$$
$$- t_p^T t_p + 2t_p^T \left[(\alpha - \beta) \left[[L_{pp} \; L_{pm} \; L_{pq}] \right] \left[\begin{bmatrix} \mathbf{1}_p \\ u_m \\ -\mathbf{1}_q \end{bmatrix} \right] + \mathbf{1}_p \right] - 2t_p^T \mathbf{1}_p$$
$$- t_m^T t_m + 2t_m^T \left[(\alpha - \beta) \left[[L_{pm}^T \; L_{mm} \; L_{mq}] \right] \left[\begin{bmatrix} \mathbf{1}_p \\ u_m \\ -\mathbf{1}_q \end{bmatrix} \right] + u_m \right]$$

$$- t_q^{\mathrm{T}} t_q + 2t_q^{\mathrm{T}} \left[(\alpha - \beta) \left[\left[L_{pq}^{\mathrm{T}} \; L_{mq}^{\mathrm{T}} \; L_{qq} \right]\right] \left[\begin{bmatrix} \mathbf{1}_p \\ u_m \\ -\mathbf{1}_q \end{bmatrix}\right] - \mathbf{1}_q \right] + 2t_q^{\mathrm{T}} \mathbf{1}_q$$

$$+ \alpha \left[\left[\mathbf{1}_p^{\mathrm{T}} \; u_m^{\mathrm{T}} \; -\mathbf{1}_q^{\mathrm{T}} \right]\right] L \left[\begin{bmatrix} \mathbf{1}_p \\ u_m \\ -\mathbf{1}_q \end{bmatrix}\right] + \mathbf{1}_p^{\mathrm{T}} \mathbf{1}_p + \mathbf{1}_q^{\mathrm{T}} \mathbf{1}_q - u_m^{\mathrm{T}} u_m.$$

Note that

$$- t_p^{\mathrm{T}} t_p + 2t_p^{\mathrm{T}} \left[(\alpha - \beta) \left[\left[L_{pp} \; L_{pm} \; L_{pq} \right]\right] \left[\begin{bmatrix} \mathbf{1}_p \\ u_m \\ -\mathbf{1}_q \end{bmatrix}\right] + \mathbf{1}_p \right]$$

$$= - \left\{ t_p - \left[(\alpha - \beta) \left[\left[L_{pp} \; L_{pm} \; L_{pq} \right]\right] \left[\begin{bmatrix} \mathbf{1}_p \\ u_m \\ -\mathbf{1}_q \end{bmatrix}\right] + \mathbf{1}_p \right] \right\}^{\mathrm{T}}$$

$$\times \left\{ t_p - \left[(\alpha - \beta) \left[\left[L_{pp} \; L_{pm} \; L_{pq} \right]\right] \left[\begin{bmatrix} \mathbf{1}_p \\ u_m \\ -\mathbf{1}_q \end{bmatrix}\right] + \mathbf{1}_p \right] \right\}$$

$$+ \left[(\alpha - \beta) \left[\left[L_{pp} \; L_{pm} \; L_{pq} \right]\right] \left[\begin{bmatrix} \mathbf{1}_p \\ u_m \\ -\mathbf{1}_q \end{bmatrix}\right] + \mathbf{1}_p \right]^{\mathrm{T}}$$

$$\times \left[(\alpha - \beta) \left[\left[L_{pp} \; L_{pm} \; L_{pq} \right]\right] \left[\begin{bmatrix} \mathbf{1}_p \\ u_m \\ -\mathbf{1}_q \end{bmatrix}\right] + \mathbf{1}_p \right].$$

Similar completion of squares for

$$- t_m^{\mathrm{T}} t_m + 2t_m^{\mathrm{T}} \left[(\alpha - \beta) \left[\left[L_{pm}^{\mathrm{T}} \; L_{mm} \; L_{mq} \right]\right] \left[\begin{bmatrix} \mathbf{1}_p \\ u_m \\ -\mathbf{1}_q \end{bmatrix}\right] + u_m \right],$$

and

$$- t_q^{\mathrm{T}} t_q + 2t_q^{\mathrm{T}} \left[(\alpha - \beta) \left[\left[L_{pq}^{\mathrm{T}} \; L_{mq}^{\mathrm{T}} \; L_{qq} \right]\right] \left[\begin{bmatrix} \mathbf{1}_p \\ u_m \\ -\mathbf{1}_q \end{bmatrix}\right] - \mathbf{1}_q \right],$$

yields

$$\Delta V = 2(t_p - \mathbf{1}_p)^{\mathrm{T}}(u_p - \mathbf{1}_p) + 2(t_q + \mathbf{1}_p)^{\mathrm{T}}(u_q + \mathbf{1}_p) \tag{8.15}$$

$$- \left\{ t_p - \left[(\alpha - \beta) \left[\left[L_{pp} \ L_{pm} \ L_{pq} \right] \right] \left[\begin{bmatrix} \mathbf{1}_p \\ u_m \\ -\mathbf{1}_q \end{bmatrix} \right] + \mathbf{1}_p \right] \right\}^{\mathrm{T}}$$

$$\times \left\{ t_p - \left[(\alpha - \beta) \left[\left[L_{pp} \ L_{pm} \ L_{pq} \right] \right] \left[\begin{bmatrix} \mathbf{1}_p \\ u_m \\ -\mathbf{1}_q \end{bmatrix} \right] + \mathbf{1}_p \right] \right\} \qquad (8.16)$$

$$- \left\{ t_m - \left[(\alpha - \beta) \left[\left[L_{pm}^{\mathrm{T}} \ L_{mm} \ L_{mq} \right] \right] \left[\begin{bmatrix} \mathbf{1}_p \\ u_m \\ -\mathbf{1}_q \end{bmatrix} \right] + u_m \right] \right\}^{\mathrm{T}}$$

$$\times \left\{ t_m - \left[(\alpha - \beta) \left[\left[L_{pm}^{\mathrm{T}} \ L_{mm} \ L_{mq} \right] \right] \left[\begin{bmatrix} \mathbf{1}_p \\ u_m \\ -\mathbf{1}_q \end{bmatrix} \right] + u_m \right] \right\} \qquad (8.17)$$

$$- \left\{ t_q - \left[(\alpha - \beta) \left[\left[L_{pq}^{\mathrm{T}} \ L_{mq}^{\mathrm{T}} \ L_{qq} \right] \right] \left[\begin{bmatrix} \mathbf{1}_p \\ u_m \\ -\mathbf{1}_q \end{bmatrix} \right] - \mathbf{1}_q \right] \right\}^{\mathrm{T}}$$

$$\times \left\{ t_q - \left[(\alpha - \beta) \left[\left[L_{pq}^{\mathrm{T}} \ L_{mq}^{\mathrm{T}} \ L_{qq} \right] \right] \left[\begin{bmatrix} \mathbf{1}_p \\ u_m \\ -\mathbf{1}_q \end{bmatrix} \right] - \mathbf{1}_q \right] \right\} \qquad (8.18)$$

$$+ s^{\mathrm{T}} \tilde{M} s, \qquad (8.19)$$

where $s = [\mathbf{1}_p; u_m; -\mathbf{1}_q]$ and $\tilde{M} = (\alpha - \beta)^2 L^2 + (3\alpha - 2\beta) L$ since $L = L^{\mathrm{T}}$. Note that the two terms in (8.15) are negative since $t_p - \mathbf{1}_p < 0$, $u_p - \mathbf{1}_p > 0$, $t_q + \mathbf{1}_q > 0$, $u_q + \mathbf{1}_q < 0$, and that the terms in (8.16), (8.17), (8.18) are all non-positive. In order to show that $\Delta V \leq 0$, it is sufficient to show that the term (8.19) is also non-positive, i.e., to show that the matrix $\tilde{M}$ is negative semi-definite.

It is easy to see that the eigenvalues of the matrix $\tilde{M}$ are $(\alpha - \beta)^2 \lambda_i^2 + (3\alpha - 2\beta)\lambda_i$, $i \in \{1, \ldots, N\}$. Hence, $\tilde{M}$ has one simple eigenvalue at zero with the corresponding right eigenvector $\mathbf{1}_N$, while all other eigenvalues are $(\alpha - \beta)^2 \lambda_i^2 + (3\alpha - 2\beta)\lambda_i$, $i \in \{2, \ldots, N\}$. We shall show that all these eigenvalues are negative. Since $\lambda_i > 0$ and $\lambda_i \leq \lambda_N$, it is sufficient to show that $\lambda_N < \frac{2\beta - 3\alpha}{(\alpha - \beta)^2}$. We note that $\lambda_N < \frac{3}{2\beta}$ from (8.10). Thus, it is sufficient to show that $\frac{3}{2\beta} < \frac{2\beta - 3\alpha}{(\alpha - \beta)^2}$. With some algebra, we see that this is equivalent to show that $\beta > \sqrt{3}\alpha$, which is true given (8.10).

Hence, we have shown that $\Delta V(y) \leq 0$. We then show that $\Delta V(y) = 0$ if and only if $y \in \mathcal{M}$. To show this, we first note that $\Delta V < 0$ if the first two terms (8.15) are not empty since they are negative. Therefore, $\Delta V = 0$ only if these terms are empty. This is the case when $|u_i| \leq 1$ for all the agents $i \in \{1, \ldots, N\}$, i.e., when the sets S_p and S_q are empty. In this case, we have

$$\Delta V = -t^\mathrm{T}t + 2t^\mathrm{T}u + 2(\alpha - \beta)t^\mathrm{T}Lu + u^\mathrm{T}(\alpha L - I)u$$

$$= -\{t - [(\alpha - \beta)L + I_N]u\}^\mathrm{T}\{t - [(\alpha - \beta)L + I_N]u\} + u^\mathrm{T}\tilde{M}u. \qquad (8.20)$$

Note that the first term in (8.20) is non-positive and is equal to zero if and only if $t = [(\alpha - \beta)L + I_N]u$.

Recall that $\tilde{M}$ has exactly one zero eigenvalue with the corresponding right eigenvector $\mathbf{1}_N$, while all other eigenvalues are negative. Therefore, the term $u^\mathrm{T}\tilde{M}u$ is also non-positive and it is equal to zero if and only if $Lu = 0$.

Hence, we conclude that $\Delta V = 0$ if and only if $t = [(\alpha - \beta)L + I_N]u$ and $Lu = 0$. Since $Lu = 0$, we obtain $t = u$. On the other hand, from (8.12), we obtain

$$t = \sigma(u(k + 1)) = u(k + 1) = u - \alpha Lv - \beta L\sigma(u) = u - \alpha Lv.$$

Thus, we see that $Lv = 0$ since $\alpha \neq 0$. Thus $v_1 = \ldots = v_N$ since the graph is connected. From (8.11), we obtain $u = -\alpha Lx - \beta Lv = -\alpha Lx$. This together with the fact that $Lu = 0$ implies that $\tilde{L}q = 0$, where the relative state $q = [q_2; \ldots; q_N]$, $q_i = x_i - x_1$ for $i \in \{2, \ldots, N\}$, and $\tilde{L}$ is given by (8.5). Since the matrix $\tilde{L}$ is non-singular, we see that $q = 0$, i.e., $x_1 = \ldots = x_N$. Therefore, $\Delta V(y) = 0$ if and only if $y \in \mathcal{M}$.

Hence, we have shown that $\Delta V(y) \leq 0$ and $\Delta V(y) = 0$ if and only if $y \in \mathcal{M}$. It then follows from a similar analysis as in the end of the proof of Theorem 8.3 that $y(k) \to \mathcal{M}$ as $k \to \infty$. Hence, the global consensus is achieved.

8.4 Simulations

In this section, we illustrate our results on the global consensus with input saturation constraints for a network with $N = 7$ agents, whose communication topology is given in Fig. 8.1.

Fig. 8.1 Network with seven agents

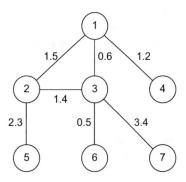

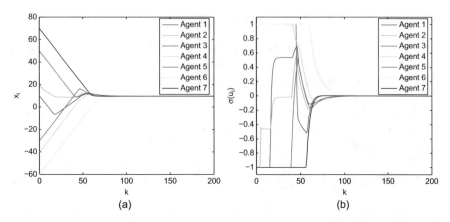

Fig. 8.2 Simulation results for neutrally stable dynamics. (**a**) The evolution of x_i. (**b**) Saturated input $\sigma(u_i)$

8.4.1 Neutrally Stable Agent Dynamics

We first consider the case of neutrally stable dynamics. In particular, single-integrator agent dynamics is considered, i.e., $\tilde{A} = 0$ and $\tilde{B} = 1$ in (8.1). We choose $\varepsilon = 0.2$. Note that $\varepsilon \in (0, \frac{2}{\lambda_N})$ since λ_N is approximately 8.7881 for the associated Laplacian matrix. Thus, the sufficient condition given in Theorem 8.3 is satisfied. The simulation results shown in Fig. 8.2 confirm the results of Theorem 8.3, i.e., consensus is still achieved even in the presence of input saturation constraints.

8.4.2 Double Integrator Agent Dynamics

We next consider the case of double integrator agent dynamics. For this case, we choose $\alpha = 0.07$ and $\beta = 0.15$ such that the sufficient condition (8.10) in Theorem 8.5 is satisfied. The simulation results for this case are given in Fig. 8.3. As can be seen, the networked dynamical system still achieves consensus even in the presence of input saturation constraints, which confirms the results of Theorem 8.5.

8.5 Literature

Most studies in the consensus literature do not consider the case where the agents are subject to input saturation. However, in almost every physical application, the actuator has bounds on its input, and thus actuator saturation is important to study. The protocol design for achieving consensus for the case with input saturation

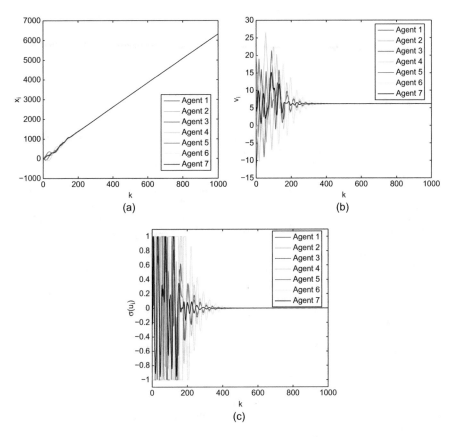

Fig. 8.3 Simulation results for double integrator dynamics. (**a**) The evolution of x_i. (**b**) The evolution of v_i. (**c**) Saturated input $\sigma(u_i)$

constraints is a challenging problem, and only few results are available. The results in this chapter are based mainly on [14]. For the single-integrator case, the authors of [3] show that any linear protocol based on the relative state information, which solves the consensus problem for the case without input saturation constraints under fixed directed communication topologies, also solves the global consensus problem in the presence of input saturation constraints. Reference [4] proposes a linear protocol based on the relative state information to solve the global consensus problem for a networked dynamical system with input saturation constraints under fixed undirected communication topologies and time-varying communication topologies. Then, it is shown in [14] that part of locally linear consensus control laws renders the global consensus for the discrete-time counterpart. Reference [13] studies semi-global regulation of output synchronization for heterogeneous networks under fixed directed communication topologies.

Acknowledgments ©2016 Elsevier. Reprinted, with permission, from Tao Yang, Ziyang Meng, Dimos V. Dimarogonas, Karl H. Johansson, "Global consensus for discrete-time multi-agent systems with input saturation constraints", Automatica, vol. 50, no. 2, 2014, pp. 499–506.

References

1. X. Bao, Z. Lin, E.D. Sontag, Finite gain stabilization of discrete-time linear systems subject to actuator saturation. Automatica **36**(2), 269–277 (2000)
2. J. Cortés, Finite-time convergence gradient flows with applications to network consensus. Automatica **42**(11), 1993–2000 (2006)
3. Y. Li, J. Xiang, W. Wei, Consensus problems for linear time-invariant multi-agent systems with saturation constraints. IET Control Theory Appl. **5**(6), 823–829 (2011)
4. Z. Meng, Z. Zhao, Z. Lin, On global leader-following consensus of identical linear dynamic systems subject to actuator saturation. Syst. Control Lett. **62**(2), 132–142 (2013)
5. A. Saberi, A.A. Stoorvogel, P. Sannuti, *Internal and External Stabilization of Linear Systems with Constraints* (Birkhäuser Basel, New York, 2012)
6. J.H. Seo, H. Shim, J. Back, Consensus of high-order linear systems using dynamic output feedback compensator: low gain approach. Automatica **45**(11), 2659–2664 (2009)
7. G. Shi, A. Saberi, A.A. Stoorvogel, On the L_p (ℓ_p) stabilization of open-loop neutrally stable linear plants with input subject to amplitude saturation. Int. J. Robust Nonlinear Control **13**(8), 735–754 (2003)
8. H. Sussmann, E. Sontag, Y. Yang, A general result on the stabilization of linear systems using bounded controls. IEEE Trans. Autom. Control **39**(12), 2411–2425 (1994)
9. A.R. Teel, Feedback Stabilization: Nonlinear Solutions to Inherently Nonlinear Problems. PhD thesis, University of California at Berkeley, Berkeley, CA, 1992
10. D. Xie, S. Wang, Consensus of second-order discrete-time multi-agent systems with fixed topology. J. Math. Anal. Appl. **387**(1), 8–16 (2012)
11. Y. Yang, E.D. Sontag, H.J. Sussmann, Global stabilization of linear discrete-time systems with bounded feedback. Syst. Control Lett. **30**(5), 273–281 (1997)
12. T. Yang, A.A. Stoorvogel, A. Saberi, Dynamic behavior of the discrete-time double integrator with saturated locally stabilizing linear state feedback laws. Int. J. Robust Nonlinear Control **23**(17), 1899–1931 (2013)
13. T. Yang, A.A. Stoorvogel, H.F. Grip et al., Semi-global regulation of output synchronization for heterogeneous networks of non-introspective, invertible agents subject to actuator saturation. Int. J. Robust Nonlinear Control **24**(3), 548–566 (2014)
14. T. Yang, Z. Meng, D.V. Dimarogonas et al., Global consensus for discrete-time multi-agent systems with input saturation constraints. Automatica **50**(2), 499–506 (2014)
15. K. You, L. Xie, Network topology and communication data rate for consensusability of discrete-time multi-agent systems. IEEE Trans. Autom. Control **56**(10), 2262–2275 (2011)
16. Y. Zhang, Y. Tian, Consentability and protocol design of multi-agent systems with stochastic switching topology. Automatica **45**(5), 1195–1201 (2009)
17. H. Zhang, F. Lewis, Z. Qu, Lyapunov, adaptive, and optimal design techniques for cooperative systems on directed communication graphs. IEEE Trans. Ind. Electron. **59**(7), 3026–3041 (2012)

Chapter 9
Control of Networked Dynamical System with Large Delays

This chapter considers synchronization of networked linear systems over networks with nonuniform time-varying delays. Since delays are inevitable in real systems and may change the stability of the original system, it is necessary and beneficial to study synchronization in the presence of delays. We focus on the case where the internal dynamics are time-varying but non-expansive (stable dynamics with a quadratic Lyapunov function). The case of directed graph is considered and we show that global asymptotic P-norm synchronization can be achieved over a directed graph with joint connectivity and arbitrarily bounded delays. Simulation results are provided to validate the theoretical results.

9.1 Problem Formulation and Control Algorithm

In this chapter, we examine the influence of the communication delays on the classical synchronization. In order to concentrate on the constraint of large delays, we specify the dynamics of (3.1) into the following equation:

$$\dot{x}_i(t) = \tilde{A}(t)x_i(t) + u_i(t), \quad i \in \mathcal{V},$$

where $\tilde{A}(t)$ is a compatible matrix. The control input for each agent is of the form

$$u_i = u_i(x_i(t), \{x_j(t - \tau_{ij}(t))\}_{j \in \mathcal{N}_i(\sigma(t))}).$$

Then, the following linear control algorithm is used:

$$u_i(t) = \gamma \sum_{j \in \mathcal{N}_i(\sigma(t))} a_{ij}(\sigma(t))(x_j(t - \tau_{ij}(t)) - x_i(t)),$$

© The Author(s), under exclusive license to Springer Nature Switzerland AG 2021
Z. Meng et al., *Modelling, Analysis, and Control of Networked Dynamical Systems*,
Systems & Control: Foundations & Applications,
https://doi.org/10.1007/978-3-030-84682-4_9

where $x_i(t) = \varphi_i(t)$, $-\tau^* \le t \le 0$, $i \in \mathcal{V}$, γ denotes the coupling gain, and $a_{ij}(p)$ is the (i, j)th entry of the adjacency matrix A_p associated with graph $\mathcal{G}_p$, for all $p \in \mathcal{P}$. Time-varying communication delay is considered, i.e., $\tau^* \ge 0$, and $\tau_{ij}(t) : \mathbb{R}_+ \to \mathbb{R}_+$, $\forall i, j \in \mathcal{V}$ is a continuous function. In addition, $\varphi_i(t) \in C([-\tau^*, 0], \mathbb{R}^n)$, $i \in \mathcal{V}$ is a vector-valued function specifying the initial state of the system, where $C([a, b], \mathbb{R}^n)$ denotes space of all real-valued continuous functions on $[a, b]$ taking values in $\mathbb{R}^n$. Then, the closed-loop system can be written as

$$\dot{x}_i(t) = \tilde{A}(t)x_i(t) + \gamma \sum_{j \in \mathcal{N}_i(\sigma(t))} a_{ij}(\sigma(t))(x_j(t - \tau_{ij}(t)) - x_i(t)), \quad i \in \mathcal{V}. \qquad (9.1)$$

We also impose the following mild assumption on communication delays.

Assumption 9.1 (Bounded Delay) *There exists a constant $\tau^* > 0$ such that for all $t \ge 0$, $\tau_{ij}(t) \le \tau^*$, for all $i, j \in \mathcal{V}$.*

Example 1: Let $\tau_{ij}(t) = i + j + \sin t$, for all $i, j \in \mathcal{V}$. It is clear that Assumption 9.1 holds for all $i, j \in \mathcal{V}$.

We focus on the case where the internal dynamics are non-expansive, i.e., the structure of $\tilde{A}(t)$ satisfying the following assumption.

Assumption 9.2 *There exists a matrix $P = P^T > 0$ such that $P\tilde{A}(t) + \tilde{A}^T(t)P \le 0$ for all $t \ge 0$.*

The objective is to establish sufficient conditions on the interaction graph with weak connectivity and the coupling gain so that synchronization is achieved. In particular, we are interested in studying global asymptotic P-norm synchronization (see Definition 4.5) and global asymptotic state synchronization (see Definition 4.2). Also note global asymptotic P-norm synchronization is a special case of global asymptotic φ-synchronization (see Definition 4.4).

9.2 Delay-Independent Gain Conditions

In this section, we focus on the case where the communication topologies are directed. Denote $x = [x_1^T, x_2^T, \ldots, x_N^T]^T \in \mathbb{R}^{Nn}$ and the initial state $x(0) = [x_1^T(0), \ldots, x_N^T(0)]^T \in \mathbb{R}^{Nn}$. Recall from Sect. 4.2, it is not hard to show that $a_* \le a_{ij}(p) \le a^*$, for all $a_{ij}(p) \ne 0$, all $i, j \in \mathcal{V}$, and all $p \in \mathcal{P}$, where a^* and a_* are defined in Sect. 4.2. The following result is established.

Theorem 9.1 *Suppose that the communication graph $\mathcal{G}_{\sigma(t)}$ is uniformly jointly strongly connected and Assumption 9.2 holds. Global asymptotic P-norm synchronization is achieved for networked dynamical system (9.1) for any $\gamma > 0$ and any $\tau_{ij}(t)$ satisfying Assumption 9.1.*

Proof Consider a Lyapunov functional $\overline{V}(x_t) = \sup_{\theta \in [-\tau^*, 0]} V(x(t + \theta))$, where x_t is a segment of the function t defined as $x_t(\theta) = x(t + \theta)$, $\theta \in [-\tau^*, 0]$, $V = \max_{i \in \mathcal{V}} V_i$, and $V_i(x_i(t)) = \|x_i\|_P^2 = x_i^T P x_i$ for all $i \in \mathcal{V}$. We first show that $D^+\overline{V}(x_t) \leq 0$, for all $t \geq 0$.

Based on the definition of $\overline{V}(x_t)$, we know that there exists a $\theta_0 \in [-\tau^*, 0]$ such that $\overline{V}(x_t) = V(x(t + \theta_0))$, where either $\theta_0 = 0$ or $\theta_0 \in [-\tau^*, 0)$. It follows from the definition of $\overline{V}(x_t)$ that $V(x(t+\theta)) \leq V(x(t+\theta_0))$ for all $\theta \in (\theta_0, 0]$ when $\theta_0 \in [-\tau^*, 0)$. Therefore, for sufficiently small $h > 0$, $\overline{V}(x_{t+h}) \leq \overline{V}(x_t)$. This implies that $D^+\overline{V}(x_t) \leq 0$ at t. When $\theta_0 = 0$, it follows that $V(x(t + \theta)) \leq V(x(t))$ for all $\theta \in [-\tau^*, 0]$ and $\overline{V}(x_t) = V(x(t))$. Let $\overline{\mathcal{V}}(t) = \{i \in \mathcal{V} : V_i(t, x(t)) = V(t, x(t))\}$ be the set of indices where the maximum is reached at time t. It then follows that

$$D^+\overline{V}(x_t) = D^+V(x(t)) = \max_{i \in \overline{\mathcal{V}}(t)} \dot{V}_i(t, x(t)) = \max_{i \in \overline{\mathcal{V}}(t)} \frac{d}{dt}\|x_i(t)\|_P^2$$

$$= \max_{i \in \overline{\mathcal{V}}(t)} \left\{ x_i^T(P\tilde{A}(t) + \tilde{A}^T(t)P)x_i + 2\gamma x_i^T(t) \right.$$

$$\left. \times P\left[\sum_{j \in \mathcal{N}_i(\sigma(t))} a_{ij}(\sigma(t))(x_j(t - \tau_{ij}(t)) - x_i(t)) \right] \right\}.$$

Therefore, we know from Assumption 9.2 that

$$D^+\overline{V}(x_t) \leq 2\gamma \max_{i \in \overline{\mathcal{V}}(t)} \sum_{j \in \mathcal{N}_i(\sigma(t))} a_{ij}(\sigma(t))(x_i^T(t)Px_j(t - \tau_{ij}(t)) - x_i^T(t)Px_i(t))$$

$$\leq \gamma \max_{i \in \overline{\mathcal{V}}(t)} \sum_{j \in \mathcal{N}_i(\sigma(t))} a_{ij}(\sigma(t))(\|x_j(t - \tau_{ij}(t))\|_P^2 - \|x_i(t)\|_P^2)$$

$$= \gamma \max_{i \in \overline{\mathcal{V}}(t)} \sum_{j \in \mathcal{N}_i(\sigma(t))} a_{ij}(\sigma(t))(V_j(x_j(t - \tau_{ij}(t))) - V_i(x_i(t)))$$

$$\leq \gamma \max_{i \in \overline{\mathcal{V}}(t)} \sum_{j \in \mathcal{N}_i(\sigma(t))} a_{ij}(\sigma(t))(V(x(t - \tau_{ij}(t))) - V(x(t)))$$

$$\leq 0,$$

where we have used the Cauchy–Schwarz inequality $x_i^T P x_j \leq \|P^{\frac{1}{2}}x_i\|\|P^{\frac{1}{2}}x_j\|$, and $V(x(t + \theta)) \leq V(x(t))$ for all $\theta \in [-\tau^*, 0]$. The second equality is due to the definition of V_j and the third inequality is due to the definition of V. To this end, we have shown that $D^+\overline{V}(x_t) \leq 0$, for all $t \geq 0$. It then follows that $\lim_{t \to \infty} \overline{V}(t) = V_\star$, where $V_\star$ is a positive constant. We also define $\overline{V}_i(x_t) = \sup_{\theta \in [-\tau^*, 0]} \|x_i(t + \theta)\|_P^2$, for all $i \in \mathcal{V}$, and next show that $\lim_{t \to \infty} \overline{V}_i(t) = V_\star$, for all $i \in \mathcal{V}$.

We shall prove the above fact by contradiction argument. It is easy to see that $\overline{V}_i(t) \leq \overline{V}(t)$, for all $i \in \mathcal{V}$ due to its definition. Therefore, for each $i \in \mathcal{V}$, there exist constants $0 \leq \chi_i \leq \psi_i \leq c_\star$, such that

$$\liminf_{t\to\infty} \overline{V}_i(t) = \chi_i, \quad \limsup_{t\to\infty} \overline{V}_i(t) = \psi_i,$$

where infimum and supremum are considered with respect to t. Suppose that there exists an agent $k_0 \in \mathcal{V}$ such that $0 \le \chi_{k_0} < \psi_{k_0} \le c_\star$.

Step 1.   Define $L_0 = \frac{\chi_{k_0}+\psi_{k_0}}{2}$. It is clear that $L_0 < c_\star$. Then there exists a time series $0 < \tilde{t}_1 < \ldots < \tilde{t}_k < \ldots$ with $\lim_{k\to\infty} \tilde{t}_k = \infty$ such that $\overline{V}_{k_0}(t) = L_0$ for all $k = 1, 2, \ldots$.

Since $\lim_{t\to\infty} \overline{V}(t) = c_\star$, it follows that for any $\varepsilon > 0$, there exists $T_1(\varepsilon) > 0$ such that for all $i \in \mathcal{V}$,

$$\overline{V}_i(t) \in [0, c_\star + \varepsilon], \quad \forall t \ge T_1(\varepsilon).$$

Let us pick up one of $\tilde{t}_k$, $k = 1, 2, \ldots$ such that it is also greater than or equal to $T_1(\varepsilon) + \tau^*$ and denote it as $\tilde{t}_{k_0}$. Therefore, $\overline{V}_{k_0}(\tilde{t}_{k_0}) = L_0$.
Denote $T^* = (N-1)T_0 + \tau^*$ with $T_0 = T + 2\tau_D$. We then give the upper bound of $V_{k_0}(t)$ agent by agent on the subintervals $t \in [(j-1)T_0, jT_0]$ for $j = 1, \ldots, N$. For all $t > \tilde{t}_{k_0}$, we have

$$\frac{d}{dt} V_{k_0}(t) \le 2\gamma x_{k_0}^{\mathrm{T}} P \sum_{j\in\mathcal{N}_{k_0}(\sigma(t))} a_{k_0 j}(\sigma(t))(x_j(t - \tau_{k_0 j}(t)) - x_{k_0}(t))$$

$$\le \gamma \sum_{j\in\mathcal{N}_{k_0}(\sigma(t))} a_{k_0 j}(\sigma(t))(V_j(x_j(t - \tau_{k_0 j}(t))) - V_{k_0}(x_{k_0}(t)))$$

$$\le \gamma a^*(N - 1)(c_\star + \varepsilon - V_{k_0}(t)).$$

It follows that for all $s \ge \tilde{t}_{k_0}$,

$$V_{k_0}(t) \le e^{-\gamma a^*(N-1)(t-s)} V_{k_0}(s) + (1 - e^{-\gamma a^*(N-1)(t-s)})(c_\star + \varepsilon). \tag{9.2}$$

From (9.2), we obtain

$$V_{k_0}(t) \le \kappa \triangleq \beta L_0 + (1 - \beta)(c_\star + \varepsilon) \tag{9.3}$$

for all $t \in [\tilde{t}_{k_0}, \tilde{t}_{k_0} + T^*]$, where $\beta = e^{-\gamma a^*(N-1)T^*}$.

Step 2.   Since $\mathcal{G}_{\sigma(t)}$ is uniformly jointly strongly connected, it is not hard to see that there exist an agent $k_1 \ne k_0 \in \mathcal{V}$ and $\hat{t}_1 \ge \tilde{t}_{k_0}$ such that $(k_0, k_1) \in \mathcal{G}_{\sigma(t)}$ for $t \in [\hat{t}_1, \hat{t}_1 + \tau_D) \subseteq [\tilde{t}_{k_0}, \tilde{t}_{k_0} + T_0)$. Based on the definition of $\hat{t}_1$, we know that $\overline{V}_{k_1}(\hat{t}_1) \in [0, V_\star + \varepsilon]$.

For $t \in [\hat{t}_1, \hat{t}_1 + \tau_D)$, there are two different scenarios, $V_{k_1}(t) > V_{k_0}(t)$ for all $t \in [\hat{t}_1, \hat{t}_1 + \tau_D)$ or there exists a time instant $\bar{t}_1 \in [\hat{t}_1, \hat{t}_1 + \tau_D)$ such that $V_{k_1}(\bar{t}_1) \le V_{k_0}(\bar{t}_1)$. We can obtain the following upper bound for both cases:

$$\frac{d}{dt}V_{k_1}(t) \le 2\gamma x_{k_1}^{\mathsf{T}} P \sum_{j \in \mathcal{N}_{k_1}(\sigma(t))} a_{k_1 j}(\sigma(t))(x_j(t - \tau_{k_1 j}(t)) - x_{k_1}(t))$$

$$\le \gamma \sum_{j \in \mathcal{N}_{k_1}(\sigma(t)) \backslash \{k_0\}} a_{k_1 j}(\sigma(t))(V_j(t - \tau_{k_1 j}(t)) - V_{k_1}(t))$$

$$+ \gamma a_{k_1 k_0}(\sigma(t))(V_{k_0}(t - \tau_{k_1 k_0}(t)) - V_{k_1}(t))$$

$$\le -\hat{\lambda}_2 V_{k_1}(t) + a_* \gamma \kappa_0 + a^*(N - 2)\gamma(c_\star + \varepsilon),$$

where $\hat{\lambda}_2 = \gamma(a^*(N - 2) + a_*)$.

Thus we obtain

$$V_{k_1}(t) \le e^{-\hat{\lambda}_2(t - \hat{t}_1)} V_{k_1}(\hat{t}_1) + \frac{[a^*(N - 2)(c_\star + \varepsilon) + a_* \kappa_0]\gamma(1 - e^{-\hat{\lambda}_2(t - \hat{t}_1)})}{\hat{\lambda}_2}.$$

Therefore, we have

$$V_{k_1}(\hat{t}_1 + \tau_D) \le \kappa_1 \overset{\triangle}{=} \alpha(c_\star + \varepsilon) + (1 - \alpha)\kappa_0, \tag{9.4}$$

where $\alpha = \frac{\hat{\lambda}_2 - a_* \gamma(1 - e^{-\hat{\lambda}_2 \tau_D})}{\hat{\lambda}_2}$. Let us now apply the same analysis as we obtained (9.3) on the agent k_1. This yields that for $t \in [\hat{t}_1 + \tau_D, \infty)$

$$V_{k_1}(t) \le e^{-(N-1)a^* \gamma(t - (\hat{t}_1 + \tau_D))} \kappa_1$$

$$+ (1 - e^{-(N-1)a^* \gamma(t - (\hat{t}_1 + \tau_D))})(c_\star + \varepsilon). \tag{9.5}$$

It then follows from (9.3), (9.4), and (9.5) that

$$V_{k_1}(t) \le \varphi_1 L_0 + (1 - \varphi_1)(c_\star + \varepsilon) \tag{9.6}$$

for all $t \in [\hat{t}_1 + \tau_D, \tilde{t}_{k_0} + T^*]$, where $\varphi_1 = (1 - \alpha)\beta^2$.

Since $0 < \varphi_1 < \beta < 1$, (9.3) and (9.6) yield

$$V_j(t) \le \varphi_1 L_0 + (1 - \varphi_1)(c_\star + \varepsilon), \quad j \in \{k_0, k_1\}, \tag{9.7}$$

for all $t \in [\hat{t}_1 + \tau_D, \tilde{t}_{k_0} + T^*]$.

Step 3. Let us next focus on the time interval $[\tilde{t}_{k_0} + T_0, \tilde{t}_{k_0} + 2T_0)$. Since $\mathcal{G}_{\sigma(t)}$ is uniformly jointly strongly connected, there exist an agent $k_2 \in \mathcal{V} \backslash \{k_0, k_1\}$ and $\hat{t}_2 \ge \tilde{t}_{k_0} + T_0$ such that there exists an edge from the set $\{k_0, k_1\}$ to the agent k_2 in $\mathcal{G}_{\sigma(t)}$ for $t \in [\hat{t}_2, \hat{t}_2 + \tau_D) \subseteq [\tilde{t}_{k_0} + T_0, \tilde{t}_{k_0} + 2T_0)$. Similar analysis as we have done for the agent k_1 will result in

$$V_{k_2}(t) \le \varphi_2 L_0 + (1 - \varphi_2)(c_\star + \varepsilon) \tag{9.8}$$

for all $t \in [\hat{t}_2 + \tau_D, \tilde{t}_{k_0} + T^*]$, where $\varphi_2 = ((1 - \alpha)\beta^2)^2$.

Since $0 < \varphi_2 < \varphi_1 < 1$, (9.7) and (9.8) yield

$$V_j(t) \le \varphi_2 L_0 + (1 - \varphi_2)(c_\star + \varepsilon), \quad j \in \{k_0, k_1, k_2\}, \tag{9.9}$$

for all $t \in [\hat{t}_2 + \tau_D, \tilde{t}_{k_0} + T^*]$.

Repeating the above procedure on time intervals $[\tilde{t}_{k_0} + 2T_0, \tilde{t}_{k_0} + 3T_0), \ldots, [\tilde{t}_{k_0} + (N - 2)T_0, \tilde{t}_{k_0} + T^*)$ yields

$$V_i(t) \le \varphi_{N-1} L_0 + (1 - \varphi_{N-1})(c_\star + \varepsilon), \quad \forall i \in \mathcal{V}$$

for all $t \in [\tilde{t}_{k_0} + (N - 1)T_0, \tilde{t}_{k_0} + T^*)$, where $\varphi_{N-1} = ((1 - \alpha)\beta^2)^{N-1}$. This implies that

$$V(t) \le \varphi_{N-1} L_0 + (1 - \varphi_{N-1})(c_\star + \varepsilon),$$

for all $t \in [\tilde{t}_{k_0} + (N - 1)T_0, \tilde{t}_{k_0} + T^*)$. Then, based on the definition of $\overline{V}(t)$, we know

$$\overline{V}(\tilde{t}_{k_0} + T^*) \le \varphi_{N-1} L_0 + (1 - \varphi_{N-1})(c_\star + \varepsilon). \tag{9.10}$$

Note that the right-hand side of (9.10) is strictly less than $c_\star$ for $\varepsilon < \frac{\varphi_{N-1}(c_\star - L_0)}{1 - \varphi_{N-1}}$. This contradicts the definition of $c_\star$. Therefore, we have shown that

$$\lim_{t \to \infty} \overline{V}_i(t) = c_\star, \quad i \in \mathcal{V}.$$

Finally, we know that

$$\lim_{t \to \infty} \|x_i(t)\|_P^2 = c_\star, \quad i \in \mathcal{V}.$$

Note that Theorem 9.1 does not prove state synchronization. We next present two further results on state synchronization.

Theorem 9.2 *Suppose that the communication graph $\mathcal{G}_{\sigma(t)}$ is uniformly jointly strongly connected and $\tilde{A}(t) = 0$. Global asymptotic state synchronization (see Definition 4.2) is achieved for the networked dynamical system (9.1) for any $\gamma > 0$ and any $\tau_{ij}(t)$ satisfying Assumption 9.1.*

Proof It is not hard to show that we can apply the proof of Theorem 9.1 to each dimension of x_i when $\tilde{A}(t) = 0$. Therefore, it follows that $\lim_{t \to \infty} \|x_{i,l}(t)\|^2 = c_{\star,l}$, $i \in \mathcal{V}$, $l \in \{1, 2, \ldots, n\}$, where $x_{i,l}$ is the lth entry of x_i. Suppose that $c_\star \ne 0$. Based on the continuity of $x_{i,l}(t)$, we know that $\lim_{t \to \infty} x_{i,l}(t) = \sqrt{c_{\star,l}}$ or $\lim_{t \to \infty} x_{i,l}(t) = -\sqrt{c_{\star,l}}$, for $i \in \mathcal{V}$. Suppose that there exists a $k \in \mathcal{V}$ such that $\lim_{t \to \infty} x_{k,l}(t) = -\sqrt{c_{\star,l}}$ and $\lim_{t \to \infty} x_{j,l}(t) = \sqrt{c_{\star,l}}$, for $j \in \mathcal{V}/\{k\}$. It follows that for any $\varepsilon_1 > 0$ and $\varepsilon_2 > 0$, there exist $\underline{T}_1(\varepsilon_1, \varepsilon_2) > 0$ such that

$x_{j,l}(t) \in [\sqrt{c_{\star,l}} - \varepsilon_2, \sqrt{c_{\star,l}} + \varepsilon_2]$ for all $j \in \mathcal{V}/\{k\}$ and $x_{k,l}(t) \in [-\sqrt{c_{\star,l}} - \varepsilon_1, -\sqrt{c_{\star,l}} + \varepsilon_1]$, for all $t \geq \underline{T}_1$. According to the fact that $\mathcal{G}_{\sigma(t)}$ is uniformly jointly strongly connected, it follows that there exists $v \in \mathcal{V}/\{k\}$ and $\underline{t}_1 \geq \underline{T}_1$ such that $(v, k) \in \mathcal{G}_{\sigma(t)}$ for $t \in [\underline{t}_1, \underline{t}_1 + \tau_D)$. Therefore, it is not hard to show that $\dot{x}_{k,l}(t) \neq 0$ and has the same sign for $t \in [\underline{t}_1, \underline{t}_1 + \tau_D)$. This further implies that $\dot{x}_{k,l}(\underline{t}_1 + \tau_D) \notin [-\sqrt{c_{\star,l}} - \varepsilon_1, -\sqrt{c_{\star,l}} + \varepsilon_1]$ by selecting $\varepsilon_1 > 0$ and $\varepsilon_2 > 0$ small enough. This indicates a contradiction and shows that $\lim_{t \to \infty} x_{i,l}(t) = d_{\star,l}$, for all $i \in \mathcal{V}$ and all $l \in \{1, 2, \ldots, N\}$, where $d_{\star,l}$ is some constant. The desired result is proven.

Theorem 9.3 *Suppose that the communication graph $\mathcal{G}_{\sigma(t)} \equiv \mathcal{G}$ is fixed and strongly connected and $\tilde{A}(t) \equiv \tilde{A}$ is time-invariant and neutrally stable. Global asymptotic state synchronization (see Definition 4.2) is achieved for the networked dynamical system (9.1) for any $\gamma > 0$ and any $\tau_{ij}(t)$ satisfying Assumption 9.1.*

Proof According to Theorem 9.1, we know that for any $\varepsilon > 0$, there exists a $T(\varepsilon) > 0$ such that $c_{\star} - \varepsilon \leq V_i(x_i(t)) \leq c_{\star} + \varepsilon$, $\forall i \in \mathcal{V}, \forall t \geq T(\varepsilon)$. Without loss of generality, we assume $c_{\star} > 0$. We use contraction argument to prove this theorem. Suppose that there exists two agents $i_0, j_0 \in \mathcal{V}$ such that $\lim \sup_{t \to \infty} \|x_{i_0}(t) - x_{j_0}(t)\|_P^2 > 0$. Therefore, there exist an infinite time sequence $t_1 < \cdots < t_\ell \ldots$ with $\lim_{\ell \to \infty} t_\ell = \infty$, and a constant $\vartheta > 0$ such that $\|x_{i_0}(t_\ell) - x_{j_0}(t_\ell)\|_P^2 = 2^{N-1}\vartheta$ for all $\ell \geq 1$. Then, following the same analysis given in the proof of Theorem 3 in [10], it follows that for any t_ℓ, there are two agents $i_*, j_* \in \mathcal{V}$ with $(i_*, j_*) \in \mathcal{E}$ such that $\|x_{i_*}(t_\ell) - x_{j_*}(t_\ell)\|_P^2 \geq \vartheta$. In the second step, since $\tilde{A}$, $x_i(t)$, $\forall i \in \mathcal{V}$ are bounded, it follows from (9.1) that $\frac{d}{dt}\|x_{j_*}(t) - x_{i_*}(t - \tau_{j_*i_*}(t))\|_P^2 \leq L^*, t \geq t_{\ell*}$, where L^* is some constant and $t_{\ell*} > T(\varepsilon) + \tau^*$. We therefore know that $\|x_{j_*}(t) - x_{i_*}(t - \tau_{j_*i_*}(t))\|_P^2 \geq \vartheta, t \in [t_{\ell*}, t_{\ell*} + \frac{\vartheta}{L^*}]$.

In the third step, it follows from (9.1) that for $t \in [t_{\ell*}, t_{\ell*} + \frac{\vartheta}{L^*}]$,

$$
\begin{aligned}
\frac{d}{dt}V_{j_*}(t) &\leq 2\gamma x_{j_*}^T P \sum_{k \in \mathcal{N}_{j_*}} a_{j_*k}(x_k(t - \tau_{j_*k}(t)) - x_{j_*}(t)) \\
&\leq 2a_{j_*i_*}\gamma x_{j_*}^T P(x_{i_*}(t - \tau_{j_*i_*}(t)) - x_{j_*}(t)) \\
&\quad + 2\gamma x_{j_*}^T P \sum_{k \in \mathcal{N}_{j_*}\setminus\{i_*\}} a_{j_*k}(x_k(t - \tau_{j_*k}(t)) - x_{j_*}(t)) \\
&\leq a_{j_*i_*}\gamma(-\|x_{j_*}(t) - x_{i_*}(t - \tau_{j_*i_*}(t))\|_P^2 \\
&\quad + \|x_{i_*}(t - \tau_{j_*i_*}(t))\|_P^2 - \|x_{j_*}\|_P^2) \\
&\quad + \gamma \sum_{k \in \mathcal{N}_{j_*}\setminus\{i_*\}} a_{j_*k}(V_k(x_k(t - \tau_{j_*k}(t))) - V_{j_*}(x_{j_*}(t))) \\
&\leq 4(N-1)a^*\varepsilon - a_*\gamma\vartheta,
\end{aligned}
$$

where we have used the fact that $2p^{\mathrm{T}}Pq = -\|p - q\|_P^2 + \|p\|_P^2 + \|q\|_P^2$, $\forall p, q \in \mathbb{R}^n$. It then follows that $V_{j_*}(t_{\ell^*} + \frac{\vartheta}{L^*}) \le c_\star + \varepsilon + (4(N - 1)a^*\varepsilon - a_*\gamma\vartheta)\frac{\vartheta}{L^*}$. Therefore, by choosing ε sufficiently small satisfying $\varepsilon < \frac{a_*\gamma\vartheta^2}{2L^*+4(N-1)a^*\vartheta}$, we have that $V_{j_*}(t_{\ell^*} + \frac{\vartheta}{L^*}) < c_\star - \varepsilon$. This indicates a contradiction and the desired result is proven.

9.3   Simulations

In this section, we use simulations to verify the theoretical results. We first consider the directed communication topology with bounded time-varying delays. We assume that $x_i \in \mathbb{R}^3$, $\gamma = 1$, $\tilde{A}(t) = \begin{bmatrix} 0 & t & 1 \\ -t & 0 & 1 \\ -1 & -1 & 0 \end{bmatrix}$, $N = 5$, and time-varying delays are $\tau_{k1}(t) = 1 + 2\sin t$, $k \in V/\{1\}$, $\tau_{k2}(t) = 2 + 2\sin t$, $k \in V/\{2\}$, $\tau_{k3}(t) = 3 + 2\sin t$, $k \in V/\{3\}$, and $\tau_{ij}(t) = 2 + \sin t$ for other $i, j \in V$. The communication topology $\mathcal{G}_\sigma(t)$ switches periodically as Fig. 9.1 at time instants $t_\ell = \ell$, $\ell = 1, 2, \ldots$.

In addition, $A_1 = \begin{bmatrix} 0 & 2 & 0 & 2 & 0 \\ 1 & 0 & 1 & 0 & 0 \\ 0 & 0 & 0 & 0 & 0 \\ 0 & 0 & 0 & 0 & 0 \\ 0 & 0 & 0 & 0 & 0 \end{bmatrix}$, $A_2 = \begin{bmatrix} 0 & 0 & 0 & 0 & 0 \\ 0 & 0 & 0 & 0 & 0 \\ 0 & 3 & 0 & 3 & 3 \\ 3 & 0 & 3 & 0 & 0 \\ 0 & 1 & 1 & 0 & 0 \end{bmatrix}$. It is not hard to check that Assumption 9.2 is satisfied and $\mathcal{G}_{\sigma(t)}$ is uniformly jointly strongly connected. Figure 9.2 shows the trajectories of x_i, $i = 1, 2, \ldots, 5$, for system (9.1). We see that all x_i, $i = 1, 2, \ldots, 5$, converge to a constant vector, which shows that global asymptotic P-norm synchronization is achieved with $P = I_3$. This agrees with the result of Theorem 9.1.

9.4   Literature

The results in this chapter are based mainly on [6]. Time delays are ubiquitous in communication networks [1] and thus synchronization for networked dynamical systems with communication time delays is also worthy of studying.

The influence of communication delays is studied in [7] and a very general result is established. It is shown that leaderless consensus is still achieved for a directed graph with arbitrarily bounded communication delays. Then, a second-

Fig. 9.1 Switching
communication topology $\mathcal{G}_1 \longrightarrow \mathcal{G}_2 \longrightarrow \mathcal{G}_1 \longrightarrow \cdots$

Fig. 9.2 State convergence for Theorem 9.1

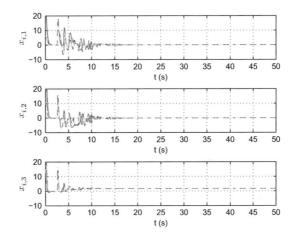

order discrete-time consensus algorithm is studied in [4] with nonuniform time delays and time-varying communication topologies. It is shown that consensus is achieved for arbitrarily bounded time delays when the velocity is dampened to zero. In addition, the authors of [2] propose a distributed protocol that allows each agent to compute the exact average of the initial values for networked dynamical systems in the presence of time-invariant delays in a finite number of time steps. An attitude synchronization problem for flexible spacecraft is considered in [3] with non-identical communication delays. The techniques of backstepping control and finite-time control are used and a non-smooth distributed algorithm is proposed. The authors of [9] establish consensus conditions for discrete-time networked dynamical systems with both input and communication delays. Critical bounds are derived using a general Nyquist stability criterion. The robustness of the linear consensus algorithm to feedback delays is studied in [8]. In particular, feedback without self-delay, feedback with identical self-delay, and feedback with different self-delays are considered, respectively. Synchronization of networked nonlinear systems is studied in [5], where both input delays and switching topologies are considered. Delay-dependent sufficient conditions are obtained for achieving local or global synchronization.

Acknowledgments ©2018 IEEE. Reprinted, with permission, from Ziyang Meng, Tao Yang, Guoqi Li, Wei Ren, Di Wu, "Synchronization of coupled dynamical systems: tolerance to weak connectivity and arbitrarily bounded time-varying delays", IEEE Transactions on Automatic Control, vol. 63, no. 6, pp. 1791–1797, 2018.

References

1. K. Aström, P. Kumar, Control: a perspective. Automatica **50**(1), 3–43 (2014)
2. T. Charalambous, Y. Yuan, T. Yang, W. Pan, C.N. Hadjicostis, M. Johansson, Distributed finite-time average consensus in digraphs in the presence of time-delays. IEEE Trans. Control Netw. Syst. **2**(4), 370–381 (2015)
3. H. Du, S. Li. Attitude synchronization for flexible spacecraft with communication delays. IEEE Trans. Autom. Control **61**, 3625–3630 (2016)
4. P. Lin, Y. Jia, Consensus of second-order discrete-time multi-agent systems with nonuniform time-delays and dynamically changing topologies. Automatica **45**, 2154–2158 (2009)
5. T. Liu, J. Zhao, D.J. Hill, Exponential synchronization of complex delayed dynamical networks with switching topology. IEEE Trans. Circuits Syst. I Regular Papers **57**(11), 2967–2980 (2010)
6. Z. Meng, W. Xia, K.H. Johansson, et al., Stability of positive switched linear systems: weak excitation and robustness to time-varying delay. IEEE Trans. Autom. Control **62**(1), 399–405 (2016)
7. L. Moreau, Stability of continuous-time distributed consensus algorithms, in *Proceedings of the IEEE Conference on Decision and Control*, pp. 3998–4003, Nassau, Bahamas, 2004
8. U. Münz, A. Papachristodoulou, F. Allgöwer, Delay robustness in consensus problems. Automatica **46**(8), 1252–1265 (2010)
9. Y.P. Tian, C.L. Liu, Consensus of multi-agent systems with diverse input and communication delays. IEEE Trans. Autom. Control **53**(9), 2122–2128 (2008)
10. T. Yang, Z. Meng, G. Shi, et al., Network synchronization with nonlinear dynamics and switching interactions. IEEE Trans. Autom. Control **61**(10), 3103–3108 (2016)

Chapter 10
Control of Networked Dynamical System with Heterogenous Dynamics

Generally speaking, cooperative control of networked dynamical systems can be classified into leaderless coordination and leader-following coordination problems depending on whether or not there is a leader specifying the global information. When the leader is dynamic, the leader-following coordination problem is specified as a coordinated tracking problem [12] and the case of a static leader falls into a special case. On the other hand, external disturbances inevitably exist in practical applications and therefore the design on disturbance rejection strategy for cooperative control of networked dynamical systems is also indispensable. In this chapter, we construct a framework to describe and study the coordinated output regulation problem for multiple heterogeneous linear systems. Here, coordinated output regulation problem is equivalent to coordinated tracking problem with disturbance rejection. Each agent is modeled as a general linear multiple-input multiple-output system with an autonomous exosystem that represents the individual offset from the group reference for the agent. The system as a whole has a group exogenous state that represents the tracking reference for the whole group. Under the constraints that the group exogenous output is only locally available to each agent and that the agents have only access to their neighbors' information, we propose observer-based feedback controllers to solve the coordinated output regulation problem using the output feedback information. A high-gain approach is used and the information interactions are allowed to be switching over a finite set of networks containing both graphs that have a directed spanning tree and graphs that do not have. Simulations are shown to validate the theoretical results.

© The Author(s), under exclusive license to Springer Nature Switzerland AG 2021   109
Z. Meng et al., *Modelling, Analysis, and Control of Networked Dynamical Systems*,
Systems & Control: Foundations & Applications,
https://doi.org/10.1007/978-3-030-84682-4_10

10.1   Problem Formulation

10.1.1   Agent Dynamics

In order to study on the influence of heterogenous dynamics of networked dynamical systems, we consider a leader–follower structure and specify the dynamics of (3.1) as N agents modeled by the linear multiple-input multiple-output (MIMO) systems:

$$\dot{x}_i = \tilde{A}_i x_i + \tilde{B}_i u_i, i \in \mathcal{V}, \tag{10.1}$$

where $x_i \in \mathbb{R}^{n_i}$ is the agent state, $u_i \in \mathbb{R}^{m_i}$ is the control input, $\tilde{A}_i \in \mathbb{R}^{n_i \times n_i}$, $\tilde{B}_i \in \mathbb{R}^{n_i \times m_i}$, and n_i and m_i are positive integers, for all $i \in \mathcal{V}$. In addition, there is an individual autonomous exosystem for each $i \in \mathcal{V}$ considered as disturbances:

$$\dot{\omega}_i = S_i \omega_i, \tag{10.2}$$

where $\omega_i \in \mathbb{R}^{q_i}$, $S_i \in \mathbb{R}^{q_i \times q_i}$, and q_i is a positive integer, for all $i \in \mathcal{V}$.

Finally, we specify the dynamics of (3.3) as a group autonomous exosystem with

$$\dot{x}_0 = \tilde{A}_0 x_0, \tag{10.3}$$

where $x_0 \in \mathbb{R}^{n_0}$, $\tilde{A}_0 \in \mathbb{R}^{n_0 \times n_0}$, and n_0 is a positive integer.

10.1.2   Available Information for Agents

For the individual autonomous exosystem tracking, the available output information for each agent $i \in \mathcal{V}$ is $y_{si} = C_{si} x_i + C_{wi} \omega_i$, where $y_{si} \in \mathbb{R}^{p_1}$, $C_{si} \in \mathbb{R}^{p_1 \times n_i}$, $C_{wi} \in \mathbb{R}^{p_1 \times q_i}$, and p_1 is a positive integer. For the group autonomous exosystem tracking, only neighbor-based output information is available due to the constrained communication. In particular, the available information is the neighbor-based sum of each agent's own output relative to that of its neighbors, i.e., $\zeta_i = \sum_{j=0}^{N} a_{ij}(\sigma(t))(y_{di} - y_{dj})$ is available for each agent $i \in \mathcal{V}$, where $a_{ij}(\sigma(t))$, $i = 1, \ldots, N$, $j = 0, 1, \ldots, N$, is entry (i, j) of the adjacency matrix $\bar{A}_{\sigma(t)}$ associated with the leader–follower graph $\bar{\mathcal{G}}_{\sigma(t)}$ at time t, $\zeta_i \in \mathbb{R}^{p_2}$, $i = 1, 2, \ldots, N$, y_{di} is represented by $y_{di} = C_{di} x_i$, $i = 1, 2, \ldots, N$ and $y_{d0} = C_0 x_0$, where $C_{di} \in \mathbb{R}^{p_2 \times n_i}$, $i = 1, 2, \ldots, N$, $C_0 \in \mathbb{R}^{p_2 \times n_0}$, $y_{di} \in \mathbb{R}^{p_2}$, $i = 0, 1, \ldots, N$, and p_2 is a positive integer. Figure 10.1 gives an example of information flow among three agents and the group exogenous input at a given time instant.

Also, the relative estimation information is available using the same communication topologies, i.e., $\hat{\zeta}_i = \sum_{j=0}^{N} a_{ij}(\sigma(t))(\hat{y}_i - \hat{y}_j)$ is available for each agent $i \in \mathcal{V}$, where $\hat{y}_i$ is an estimate produced internally by each agent $i \in \mathcal{V}$, $\hat{\zeta}_i \in \mathbb{R}^{p_2}$,

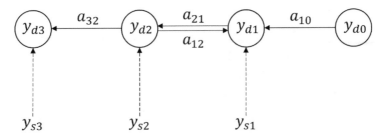

Fig. 10.1 Information flow associated with three agents 1, 2, 3, the exogenous inputs $\omega_1, \omega_2, \omega_3$, and the group exogenous input 0

$i = 1, 2, \ldots, N$, and $\widehat{y}_i \in \mathbb{R}^{p_2}$, $i = 0, 1, \ldots, N$, which will be given explicitly in Sect. 10.2.

10.1.3 Switching Topologies

For the communication topology set $\{\bar{\mathcal{G}}_p\}_{p \in \mathcal{P}}$, we assume that $\bar{\mathcal{G}}_p$, $\forall p \in \mathcal{P}_c$, is leader rooted (see Definition 3.2). Without loss of generality, we relabel $\mathcal{P}_c :=$ $\{1, 2, \ldots, \varrho_1\}$, $1 \leq \varrho_1 \leq \varrho$. The remaining graphs are labeled as $\bar{\mathcal{G}}_p$, $\forall p \in \mathcal{P}_d$, where $\mathcal{P}_d := \{\varrho_1 + 1, \varrho_1 + 2, \ldots, \varrho\}$. Denote the graph set $\bar{\mathbb{G}}_c = \{\bar{\mathcal{G}}_p\}_{p \in \mathcal{P}_c}$ and the graph set $\bar{\mathbb{G}}_d = \{\bar{\mathcal{G}}_p\}_{p \in \mathcal{P}_d}$, respectively. We also denote $T_{\bar{t}_0}^d(t)$ and $T_{\bar{t}_0}^c(t)$ the total activation time when $\bar{\mathcal{G}}_{\sigma(\varsigma)} \in \bar{\mathbb{G}}_d$ and total activation time when $\bar{\mathcal{G}}_{\sigma(\varsigma)} \in \bar{\mathbb{G}}_c$, respectively, during $\varsigma \in [\bar{t}_0, t)$ for $\bar{t}_0 \geq 0$.

Assumption 10.1 *There exist positive constants κ and $\bar{t}_0 \geq 0$ such that $T_{\bar{t}_0}^c(t) \geq \kappa T_{\bar{t}_0}^d(t)$ for all $t \geq \bar{t}_0$.*

10.1.4 Control Objective and Control Architecture

The control objective of each agent is to track a given trajectory determined by the combination of the group references x_0 and the individual offset ω_i, $i = 1, 2, \ldots, N$. Such a combination is captured by the coordinated output regulation tracking error (i.e., the total tracking error representing the combination of both individual tracking and group tracking of each agent):

$$e_i = D_{si} x_i + D_{wi} \omega_i + D_0 x_0, , \tag{10.4}$$

where $D_{si} \in \mathbb{R}^{p_3 \times n_i}$, $D_{wi} \in \mathbb{R}^{p_3 \times q_i}$, $e_i \in \mathbb{R}^{p_3}$, $i \in \mathcal{V}$, $D_0 \in \mathbb{R}^{p_3 \times n_0}$, and p_3 is a positive integer. Thus, our objective is to guarantee that $\lim_{t \to \infty} e_i(t) = 0$ for all

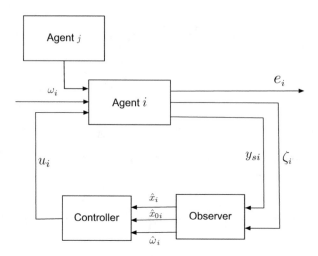

Fig. 10.2 Control architecture for agent i

$i \in \mathcal{V}$. One example of the overall control can correspond to a formation control problem, where ω_i encodes the relative position between each agent and the leader, while the leader x_0 defines the overall motion of the group.

Our goal is to design an observer-based controller with the available individual output information and the neighbor-based group output information to solve this problem. The control of each agent is supposed to have the structure depicted in Fig. 10.2. In the next section, we will specify the design procedure.

10.2   A Unified Observer-Based Control Algorithm

As suggested by Fig. 10.2, the design procedure to solve the coordinated output regulation problem includes two main steps: the first one is the state feedback control design and the second one is the observer design for the group autonomous exosystem, the individual autonomous exosystem, and the internal state information for each agent.

10.2.1   Redundant Modes

Before designing the state feedback control and distributed observer, we need first to remove the redundant modes that have no effect on y_{si} and $y_{di} - y_{d0}$. We impose the following assumption on the structure of the systems.

Assumption 10.2 *For each* $i \in \mathcal{V}$, $\left(\tilde{A}_i, \begin{bmatrix} C_{si} \\ C_{di} \end{bmatrix}\right)$ *is observable,* (S_i, C_{wi}) *is observable, and* $(\tilde{A}_0, C_0)$ *is observable.*

We then write the state and output of each agent in the compact form: $\begin{bmatrix} \dot{x}_i \\ \dot{\omega}_i \\ \dot{x}_0 \end{bmatrix} =$

$$\begin{bmatrix} \tilde{A}_i & 0 & 0 \\ 0 & S_i & 0 \\ 0 & 0 & \tilde{A}_0 \end{bmatrix} \begin{bmatrix} x_i \\ \omega_i \\ x_0 \end{bmatrix} + \begin{bmatrix} \tilde{B}_i \\ 0 \\ 0 \end{bmatrix} u_i, \text{ and } \begin{bmatrix} y_{si} \\ y_{di} - y_{d0} \end{bmatrix} = \begin{bmatrix} C_{si} & C_{wi} & 0 \\ C_{di} & 0 & -C_0 \end{bmatrix} \begin{bmatrix} x_i \\ \omega_i \\ x_0 \end{bmatrix}.$$

Given that Assumption 10.2 is satisfied, we can perform the state transformation given in Step 1 of [3] by considering ω_i and x_0 together. We construct a new state $\bar{x}_i = W_i \begin{bmatrix} x_i \\ \omega_i \\ x_0 \end{bmatrix}$ with the dynamics

$$\dot{\bar{x}}_i = \overline{A}_i \bar{x}_i + \overline{B}_i u_i = \begin{bmatrix} \tilde{A}_i & \overline{A}_{i12} \\ 0 & \overline{A}_{i22} \end{bmatrix} \bar{x}_i + \begin{bmatrix} \tilde{B}_i \\ 0 \end{bmatrix} u_i, \tag{10.5a}$$

$$\begin{bmatrix} y_{si} \\ e_{di} \end{bmatrix} = \overline{C}_i \bar{x}_i = \begin{bmatrix} C_{si} & \overline{C}_{i21} \\ C_{di} & \overline{C}_{i22} \end{bmatrix} \bar{x}_i, \tag{10.5b}$$

where $e_{di} = y_{di} - y_{d0}$, and the details of $W_i, \overline{A}_i, \overline{B}_i, \overline{C}_i$ can be found in [3]. It was shown that the pair $(\overline{A}_i, \overline{C}_i)$ is observable and the eigenvalues of $\overline{A}_{i22}$ are a subset of the eigenvalues of S_i and $\tilde{A}_0$, $i = 1, 2, \ldots, N$.

10.2.2 Regulated State Feedback Control Law

We now design a controller to regulate e_i to zero for each agent based on the state information $\bar{x}_i = [\bar{x}_{i1}^T, \bar{x}_{i2}^T]^T$, where $\bar{x}_{i1} \in \mathbb{R}^{n_i}$. We impose the following assumption on the structure of the systems.

Assumption 10.3 *For each* $i \in \mathcal{V}$, $(\tilde{A}_i, \tilde{B}_i)$ *is stabilizable,* $(\tilde{A}_i, \tilde{B}_i, D_{si})$ *is right-invertible, and* $(\tilde{A}_i, \tilde{B}_i, D_{si})$ *has no invariant zeros in the closed right-half complex plane that coincide with the eigenvalues of* S_i *or* $\tilde{A}_0$.

Lemma 10.1 *Let Assumption 10.3 hold. Then, the regulator equations (10.6) are solvable and the state feedback controller* $u_i = F_i(\bar{x}_{i1} - \Pi_i \bar{x}_{i2}) + \Gamma_i \bar{x}_{i2}$ *guarantees that* $\lim_{t \to \infty} e_i(t) = 0$ *for* $i \in \mathcal{V}$, *where* Π_i, Γ_i *are the solutions of the equations*

$$\Pi_i \overline{A}_{i22} = \tilde{A}_i \Pi_i + \overline{A}_{i12} + \tilde{B}_i \Gamma_i, \tag{10.6a}$$

$$0 = D_{si} \Pi_i + \begin{bmatrix} D_{wi} & D_0 \end{bmatrix}, \quad i \in \mathcal{V}, \tag{10.6b}$$

and F_i is chosen such that $\tilde{A}_i + \tilde{B}_i F_i$ is Hurwitz.

Proof Following from Corollary 2.5.1 of [13] and a similar analysis as in the proof of Lemma 3 in [3], we can show that the regulator equations (10.6) are solvable given that Assumption 10.3 is satisfied. Then, by considering $\dot{\overline{x}}_{i2} = \overline{A}_{i22}\overline{x}_{i2}$ as the exosystem and $\dot{x}_i = \tilde{A}_i x_i + \tilde{B}_i u_i$ as the system to be regulated from the classic output regulation result [2], we know that $u_i = F_i(\overline{x}_{i1} - \Pi_i \overline{x}_{i2}) + \Gamma_i \overline{x}_{i2}$ guarantees that $\lim_{t\to\infty} e_i(t) = 0$ for all $i \in \mathcal{V}$, where Π_i and Γ_i are the solutions of the regulator equations (10.6).

We next design an observer to estimate $\overline{x}_i$ based on the output information y_{si} and ζ_i for each agent.

10.2.3   Pseudo-Identical Linear Transformation

Note that the individual offset ω_i can be estimated from y_{si} and the group reference x_0 can be estimated from $\widehat{\zeta}_i$. In contrast, the internal state information x_i for each agent can be obtained from either y_{si} or $\widehat{\zeta}_i$. In this section, we use the combination of y_{si} and $\widehat{\zeta}_i$ to develop a unified observer design.

We define $\chi_i = T_i \overline{x}_i \in \mathbb{R}^{p\overline{n}}$, $i = 1, 2, \ldots, N$, where $\overline{n} = n_0 + \max_{i\in\mathcal{V}}(n_i + q_i)$,

$p = p_1 + p_2$, and $T_i = \begin{bmatrix} \overline{C}_i \\ \vdots \\ \overline{C}_i \overline{A}_i^{\overline{n}-1} \end{bmatrix}$. Note that T_i is full column rank since the pair

$(\overline{A}_i, \overline{C}_i)$, $i = 1, 2, \ldots, N$, is observable. This implies that $T_i^{\mathsf{T}} T_i$ is non-singular. Therefore, from (10.5) and the above state transformation, we obtain

$$\dot{\chi}_i = (\mathbf{A} + \mathbf{L}_i)\chi_i + \mathbf{B}_i u_i, \tag{10.7a}$$

$$\begin{bmatrix} y_{si} \\ e_{di} \end{bmatrix} = \mathbf{C}\chi_i, \quad i \in \mathcal{V}, \tag{10.7b}$$

where $\mathbf{A} = \begin{bmatrix} 0 & I_{p(\overline{n}-1)} \\ 0 & 0 \end{bmatrix} \in \mathbb{R}^{p\overline{n}\times p\overline{n}}$, $\mathbf{L}_i = \begin{bmatrix} 0 \\ l_i \end{bmatrix}$, $\mathbf{B}_i = T_i \overline{B}_i$, $\mathbf{C} = \begin{bmatrix} I_p & 0 \end{bmatrix} \in \mathbb{R}^{p\times p\overline{n}}$ for some matrix l_i.

10.2.4   Unified Observer Design

Motivated by [3], based on the available output information y_{si} and the neighbor-based group output information ζ_i, the distributed observer for (10.7) is proposed to be

$$\dot{\widehat{\chi}}_i = (\mathbf{A} + \mathbf{L}_i)\widehat{\chi}_i + \mathbf{B}_i u_i + S(\varepsilon)\mathbf{P}\mathbf{C}^{\mathrm{T}}$$

$$\times \left(\begin{bmatrix} y_{si} \\ \sum_{j=0}^{N} a_{ij}(\sigma(t))(y_{di} - y_{dj}) \end{bmatrix} - \begin{bmatrix} \widehat{y}_{si} \\ \sum_{j=0}^{N} a_{ij}(\sigma(t))(\widehat{y}_i - \widehat{y}_j) \end{bmatrix} \right), \ i \in \mathcal{V},$$

$$\tag{10.8}$$

where $a_{ij}(\sigma(t))$, $i = 1, \ldots, N$, $j = 0, 1, \ldots, N$, is the entry (i, j) of the adjacency matrix $\bar{A}_{\sigma(t)}$ associated with $\bar{\mathcal{G}}_{\sigma(t)}$,

$$\widehat{y}_{si} = \mathbf{C}_1 \widehat{\chi}_i, \ i \in \mathcal{V}, \tag{10.9}$$

$$\widehat{y}_i = \mathbf{C}_2 \widehat{\chi}_i, \ i \in \mathcal{V}, \tag{10.10}$$

$\mathbf{C}_1$ is first p_1 rows of $\mathbf{C}$, $\mathbf{C}_2$ is the remaining p_2 rows of $\mathbf{C}$, and $\widehat{y}_0 = 0$. In addition, $S(\varepsilon) = \mathrm{diag}(I_p \varepsilon^{-1}, I_p \varepsilon^{-2}, \ldots, I_p \varepsilon^{-\bar{n}})$, where $\varepsilon \in (0, 1]$ is a positive constant to be determined, and $\mathbf{P} = \mathbf{P}^{\mathrm{T}}$ is a positive definite matrix satisfying

$$\mathbf{AP} + \mathbf{PA}^{\mathrm{T}} - 2\mathbf{PC}^{\mathrm{T}} \begin{bmatrix} I_{p_1} & 0 \\ 0 & \theta I_{p_2} \end{bmatrix} \mathbf{CP} + I_{p\bar{n}} = 0, \tag{10.11}$$

where $\theta = \min_{p \in \mathcal{P}_c} \beta_p$, β_p is a positive constant satisfying $\beta_p < \min \mathrm{Re}\{\lambda(\bar{L}_p)\}$, $p \in \mathcal{P}_c$, and $\min \mathrm{Re}\{\lambda(\bar{L}_p)\}$ denote the minimum value of all the real parts of the eigenvalues of $\bar{L}_p$. Note that the existence of $\mathbf{P}$ is due to the fact that $\left(\mathbf{A}, \begin{bmatrix} I_{p_1} & 0 \\ 0 & \sqrt{\theta} I_{p_2} \end{bmatrix} \mathbf{C} \right)$ is observable.

Lemma 10.2

- *All the eigenvalues of $\bar{L}_p$ are in the closed right-half plane and those on the imaginary axis are simple, where $\bar{L}_p$ is associated with $\bar{\mathcal{G}}_p$, $p \in \mathcal{P}$.*
- *Furthermore, all the eigenvalues of $\bar{L}_p$ are in the open right-half plane for $p \in \mathcal{P}_c$.*

Proof See Lemma 3.3 in Sect. 2.3.

Lemma 10.3 *Let Assumptions 10.1–10.2 hold and assume that $\kappa \geq \frac{\alpha + 4 \max\{\theta, 1\} \lambda_{\max}^2(\mathbf{P})}{1 - \alpha}$, where $\alpha \in (0, 1)$, θ, and $\mathbf{P}$ are given by (10.11). Then, there exists an $\varepsilon^* \in (0, 1]$ such that, if $\varepsilon \in (0, \varepsilon^*]$, $\lim_{t \to \infty}(\chi_i(t) - \widehat{\chi}_i(t)) = 0$ for all $i \in \mathcal{V}$, for system (10.8).*

Proof Note that for all $i = 1, 2, \ldots, N$, $\sum_{j=0}^{N} a_{ij}(\sigma(t))(y_{di} - y_{dj}) = \sum_{j=1}^{N} l_{ij}(\sigma(t))(y_{dj} - y_{d0}) = \sum_{j=1}^{N} l_{ij}(\sigma(t)) e_{dj}$. Define $\widetilde{\chi}_i = \chi_i - \widehat{\chi}_i$. It then follows from (10.7) and (10.8) that for all $i \in \mathcal{V}$,

$$\dot{\tilde{\chi}}_i = (\mathbf{A} + \mathbf{L}_i)\tilde{\chi}_i - S(\varepsilon)\mathbf{P}\mathbf{C}^{\mathrm{T}} \left[\begin{array}{c} y_{si} - \widehat{y}_{si} \\ \sum_{j=1}^{N} l_{ij}(\sigma(t))(e_{dj} - \widehat{y}_j) \end{array} \right],$$

where $l_{ij}(\sigma(t))$, $i = 1, \ldots, N$, $j = 1, \ldots, N$, is the (i, j)-th entry of the adjacency matrix $\bar{L}_{\sigma(t)}$ associated with $\bar{\mathcal{G}}_{\sigma(t)}$. It follows that $\dot{\tilde{\chi}}_i = (\mathbf{A} + \mathbf{L}_i)\tilde{\chi}_i - S(\varepsilon)\mathbf{P}\mathbf{C}^{\mathrm{T}} \left[\begin{array}{c} \mathbf{C}_1 \tilde{\chi}_i \\ \mathbf{C}_2 \sum_{j=1}^{n} l_{ij}(\sigma(t))\tilde{\chi}_j \end{array} \right]$, $i = 1, 2 \ldots, N$. By introducing $\xi_i = \varepsilon^{-1} S^{-1}(\varepsilon)\tilde{\chi}_i$ and after some manipulations, we have that $\varepsilon \dot{\xi}_i = (\mathbf{A} + \mathbf{L}_{i\varepsilon})\xi_i - \mathbf{P}\mathbf{C}^{\mathrm{T}} \left[\mathbf{C}_1 \xi_i \mathbf{C}_2 \sum_{j=1}^{N} l_{ij}(\sigma(t))\xi_j \right]$, $i = 1, 2 \ldots, N$, where

$$\mathbf{L}_{i\varepsilon} = \left[\begin{array}{c} 0 \\ \varepsilon^{\bar{n}+1} \bar{L}_i S(\varepsilon) \end{array} \right] = O(\varepsilon).$$

Note that $\left[\begin{array}{c} \mathbf{C}_1 \xi_i \\ \mathbf{C}_2 \xi_i \end{array} \right] = \mathbf{C}\xi_i$, for all $i \in \mathcal{V}$. The overall dynamics can be written as

$$\varepsilon \dot{\xi} = \left(I_N \otimes \mathbf{A} + \mathbf{L}_\varepsilon - (I_N \otimes \mathbf{P}\mathbf{C}^{\mathrm{T}}) \right)$$
$$\times \left(I_N \otimes \left[\begin{array}{cc} I_{p_1} & 0 \\ 0 & 0 \end{array} \right] + \bar{L}_\sigma \otimes \left[\begin{array}{cc} 0 & 0 \\ 0 & I_{p_2} \end{array} \right] \right) (I_N \otimes \mathbf{C}) \right) \xi, \tag{10.12}$$

where $\xi = [\xi_1^{\mathrm{T}}, \xi_2^{\mathrm{T}}, \ldots, \xi_N^{\mathrm{T}}]^{\mathrm{T}}$ and $\mathbf{L}_\varepsilon = \mathrm{diag}(\mathbf{L}_{1\varepsilon}, \mathbf{L}_{2\varepsilon}, \ldots, \mathbf{L}_{N\varepsilon})$.

Note that $-\bar{L}_p$, $p \in \mathcal{P}_c$ is a Hurwitz matrix according to Lemma 10.2. Therefore, we can always guarantee that $-\bar{L}_p + \beta_p I_N$ is also a Hurwitz matrix by choosing β_p sufficiently small. In particular, we choose β_p as a positive constant satisfying $\beta_p < \min \mathrm{Re}\{\lambda(\bar{L}_p)\}$, $p \in \mathcal{P}_c$. Then, we define the piecewise Lyapunov function candidate $V_p = \varepsilon \xi^{\mathrm{T}}(P_p \otimes \mathbf{P}^{-1})\xi$, where P_p is a positive definite matrix satisfying

$$P_p(-\bar{L}_p + \beta_p I_N) + (-\bar{L}_p + \beta_p I_N)^{\mathrm{T}} P_p = -I_N < 0, \quad p \in \mathcal{P}_c,$$

$$P_p(-\bar{L}_p) + (-\bar{L}_p)^{\mathrm{T}} P_p \leq 0, \qquad p \in \mathcal{P}_d,$$

where the second inequality is due to Lemma 10.2.

It then follows that for all $p \in \mathcal{P}_c$,

$$\dot{V}_p \leq 2\xi^{\mathrm{T}} \left(P_p \otimes \mathbf{P}^{-1}\mathbf{A} \right) \xi + 2\xi^{\mathrm{T}} \left(P_p \otimes \mathbf{P}^{-1} \right) \mathbf{L}_\varepsilon \xi$$

$$- 2\xi^{\mathrm{T}} \left(P_p \otimes \left(\mathbf{C}^{\mathrm{T}} \left[\begin{array}{cc} I_{p_1} & 0 \\ 0 & 0 \end{array} \right] \mathbf{C} \right) \right) \xi - 2\xi^{\mathrm{T}} \left(P_p \bar{L}_p \otimes \left(\mathbf{C}^{\mathrm{T}} \left[\begin{array}{cc} 0 & 0 \\ 0 & I_{p_2} \end{array} \right] \mathbf{C} \right) \right) \xi$$

$$\leq \xi^{\mathrm{T}} \left(P_p \otimes \left(\mathbf{P}^{-1}\mathbf{A} + \mathbf{A}^{\mathrm{T}}\mathbf{P}^{-1} - 2\theta \mathbf{C}^{\mathrm{T}} \left[\begin{array}{cc} 0 & 0 \\ 0 & I_{p_2} \end{array} \right] \mathbf{C} - 2\mathbf{C}^{\mathrm{T}} \left[\begin{array}{cc} I_{p_1} & 0 \\ 0 & 0 \end{array} \right] \mathbf{C} \right) \right) \xi$$

$$+ 2\xi^{\mathrm{T}}\left(P_p \otimes \mathbf{P}^{-1}\right)\mathbf{L}_\varepsilon\xi - \xi^{\mathrm{T}}\left(\left(2P_p\bar{L}_p - 2\theta P_p\right) \otimes (\mathbf{C}^{\mathrm{T}}\begin{bmatrix} 0 & 0 \\ 0 & I_{p2} \end{bmatrix}\mathbf{C})\right)\xi$$

$$\leq \xi^{\mathrm{T}}\left(P_p \otimes \left(\mathbf{P}^{-1}\left(\mathbf{A}\mathbf{P} + \mathbf{P}\mathbf{A}^{\mathrm{T}} - 2\mathbf{P}\mathbf{C}^{\mathrm{T}}\begin{bmatrix} I_{p1} & 0 \\ 0 & \theta I_{p2} \end{bmatrix}\mathbf{C}\mathbf{P}\right)\mathbf{P}^{-1}\right)\right)\xi$$

$$- \xi^{\mathrm{T}}\left(\left(P_p\bar{L}_p + \bar{L}_p^{\mathrm{T}}P_p - 2\beta_p P_p\right) \otimes (\mathbf{C}^{\mathrm{T}}\begin{bmatrix} 0 & 0 \\ 0 & I_{p2} \end{bmatrix}\mathbf{C})\right)\xi$$

$$+ 2\lambda_{\max}(P_p)\lambda_{\max}(\mathbf{P}^{-1})\|\mathbf{L}_\varepsilon\|\|\xi\|^2$$

$$\leq -\xi^{\mathrm{T}}\left(P_p \otimes (\mathbf{P}^{-1}\mathbf{P}^{-1})\right)\xi - \xi^{\mathrm{T}}\left(I_N \otimes (\mathbf{C}^{\mathrm{T}}\begin{bmatrix} 0 & 0 \\ 0 & I_{p2} \end{bmatrix}\mathbf{C})\right)\xi$$

$$+ \frac{2\lambda_{\max}(P_p)\lambda_{\max}(\mathbf{P}^{-1})\|\mathbf{L}_\varepsilon\|}{\varepsilon\lambda_{\min}(P_p)\lambda_{\min}(\mathbf{P}^{-1})}V_p$$

$$\leq -\xi^{\mathrm{T}}\left(P_p \otimes (\mathbf{P}^{-1}\mathbf{P}^{-1})\right)\xi + \frac{2\lambda_{\max}(P_p)\lambda_{\max}(\mathbf{P}^{-1})\|\mathbf{L}_\varepsilon\|}{\varepsilon\lambda_{\min}(P_p)\lambda_{\min}(\mathbf{P}^{-1})}V_p$$

$$\leq -\left(\frac{\lambda_{\min}(\mathbf{P}^{-1})}{\varepsilon} - \frac{2\lambda_{\max}(P_p)\lambda_{\max}(\mathbf{P}^{-1})\|\mathbf{L}_\varepsilon\|}{\varepsilon\lambda_{\min}(P_p)\lambda_{\min}(\mathbf{P}^{-1})}\right)V_p,$$

where we have used (10.11) and the fact that $\theta \leq \beta_p$, $p \in \mathcal{P}_c$. It then follows that $\dot{V}_p \leq -\frac{1}{\varepsilon}\lambda^c V_p, \forall p \in \mathcal{P}_c$, if $\|\mathbf{L}_\varepsilon\| < \frac{\lambda_{\min}(P_p)\lambda_{\min}(\mathbf{P})}{4\lambda_{\max}(P_p)\lambda_{\max}^2(\mathbf{P})}$, where $\lambda^c = \frac{1}{2\lambda_{\max}(\mathbf{P})}$.

On the other hand, for all $p \in \mathcal{P}_d$, we have that

$$\dot{V}_p \leq 2\xi^{\mathrm{T}}\left(P_p \otimes (\mathbf{P}^{-1}\mathbf{A})\right)\xi + 2\xi^{\mathrm{T}}\left(P_p \otimes \mathbf{P}^{-1}\right)\mathbf{L}_\varepsilon\xi$$

$$- 2\xi^{\mathrm{T}}\left(P_p \otimes (\mathbf{C}^{\mathrm{T}}\begin{bmatrix} I_{p1} & 0 \\ 0 & 0 \end{bmatrix}\mathbf{C})\right)\xi - 2\xi^{\mathrm{T}}\left(P_p\bar{L}_p \otimes (\mathbf{C}^{\mathrm{T}}\begin{bmatrix} 0 & 0 \\ 0 & I_{p2} \end{bmatrix}\mathbf{C})\right)\xi$$

$$\leq \xi^{\mathrm{T}}\left(P_p \otimes (\mathbf{P}^{-1}(\mathbf{A}\mathbf{P} + \mathbf{P}\mathbf{A}^{\mathrm{T}})\mathbf{P}^{-1})\right)\xi + 2\lambda_{\max}(P_p)\lambda_{\max}(\mathbf{P}^{-1})\|\mathbf{L}_\varepsilon\|\|\xi\|^2$$

$$\leq 2\xi^{\mathrm{T}}\left(P_p \otimes (\mathbf{C}^{\mathrm{T}}\begin{bmatrix} I_{p1} & 0 \\ 0 & \theta I_{p2} \end{bmatrix}\mathbf{C})\right)\xi - \frac{\lambda_{\min}(\mathbf{P}^{-1})}{\varepsilon}V_p$$

$$+ \frac{2\lambda_{\max}(P_p)\lambda_{\max}(\mathbf{P}^{-1})\|\mathbf{L}_\varepsilon\|}{\varepsilon\lambda_{\min}(P_p)\lambda_{\min}(\mathbf{P}^{-1})}V_p,$$

where we have used (10.11). Note that $\lambda_{\max}\left(\mathbf{C}^{\mathrm{T}}\begin{bmatrix} I_{p1} & 0 \\ 0 & \theta I_{p2} \end{bmatrix}\mathbf{C}\right) = \max\{\theta, 1\}$. It follows that $\dot{V}_p \leq \frac{1}{\varepsilon}\lambda^d V_p, \forall p \in \mathcal{P}_d$, if $\|\mathbf{L}_\varepsilon\| < \frac{\lambda_{\min}(P_p)\lambda_{\min}(\mathbf{P})}{2\lambda_{\max}(P_p)\lambda_{\max}^2(\mathbf{P})}$, where $\lambda^d = 2\max\{\theta, 1\}\lambda_{\max}(\mathbf{P})$.

Following a similar analysis as that in [7, 23], we let $\sigma = p$ on $[t_{j-1}, t_j)$ for $p \in \mathcal{P}$. Then, for any t satisfying $0 < t_1 < \cdots < t_\ell < t < t_{\ell+1}$, define $V = \varepsilon \xi^{\mathrm{T}}(P_{\sigma(t)} \otimes \mathbf{P}^{-1})\xi$ for (10.12). We have that, $\forall \zeta \in [t_{j-1}, t_j)$,

$$V(\zeta) \leq e^{-\frac{1}{\varepsilon}\lambda^c(\zeta - t_{j-1})} V(t_{j-1}), \quad p \in \mathcal{P}_c,$$

$$V(\zeta) \leq e^{\frac{1}{\varepsilon}\lambda^d(\zeta - t_{j-1})} V(t_{j-1}), \quad p \in \mathcal{P}_d.$$

Define $a = \frac{\lambda_{\max}(\mathbf{P})}{\lambda_{\min}(\mathbf{P})} \max_{p,q \in \mathcal{P}} \frac{\lambda_{\max}(P_p)}{\lambda_{\min}(P_q)}$. We then know that $V(t_j) \leq a \lim_{t \uparrow t_j} V(t)$. Thus, it follows that $V(t) \leq a^\rho e^{\frac{1}{\varepsilon}\lambda^d T^d_{\bar{t}_0}(t) - \frac{1}{\varepsilon}\lambda^c T^c_{\bar{t}_0}(t)} V(\bar{t}_0)$, where ρ denotes times of switching during $[\bar{t}_0, t)$. Note that $\rho \leq \frac{t - \bar{t}_0}{\tau_D}$. Given that $\kappa \geq \kappa^* = \frac{\lambda^d + \lambda}{\lambda^c - \lambda}$, for some $\lambda \in (0, \lambda^c)$, it follows from Assumption 10.1 that $T^c_{\bar{t}_0}(t) \geq \kappa^* T^d_{\bar{t}_0}(t)$ for all $t \geq \bar{t}_0$. This implies that $\lambda^d T^d_{\bar{t}_0}(t) - \lambda^c T^c_{\bar{t}_0}(t) \leq -\lambda(T^d_{\bar{t}_0}(t) + T^c_{\bar{t}_0}(t))$, for all $t \geq \bar{t}_0$, and we therefore know that

$$V(t) \leq a^\rho e^{-\frac{1}{\varepsilon}\lambda(t - \bar{t}_0)} V(\bar{t}_0) \leq e^{\frac{t - \bar{t}_0}{\tau_D} \ln a - \frac{1}{\varepsilon}\lambda(t - \bar{t}_0)} V(\bar{t}_0) = e^{-\left(\frac{1}{\varepsilon}\lambda - \frac{\ln a}{\tau_D}\right)(t - \bar{t}_0)} V(\bar{t}_0).$$

Furthermore, set $\lambda = \alpha \lambda^c$, where some $\alpha \in (0, 1)$. We then have that $\kappa^* = \frac{\alpha + 4 \max\{\theta, 1\}\lambda^2_{\max}(\mathbf{P})}{1 - \alpha}$, and

$$V(t) \leq e^{-\left(\frac{\alpha}{2\varepsilon\lambda_{\max}(\mathbf{P})} - \frac{\ln a}{\tau_D}\right)(t - \bar{t}_0)} V(\bar{t}_0).$$

It follows that if $\varepsilon < \frac{\alpha \tau_D}{2\lambda_{\max}(\mathbf{P}) \ln a}$, we have for (10.12) that $\|\xi(t)\| \leq c^* e^{-\frac{1}{2}\left(\frac{\alpha}{2\varepsilon\lambda_{\max}(\mathbf{P})} - \frac{\ln a}{\tau_D}\right)(t - \bar{t}_0)} \|\xi(\bar{t}_0)\|$, where $c^* = \sqrt{\frac{\lambda_{\max}(\mathbf{P}) \max_{p \in \mathcal{P}} \lambda_{\max}(P_p)}{\lambda_{\min}(\mathbf{P}) \min_{p \in \mathcal{P}} \lambda_{\min}(P_p)}}$.

Therefore, we choose ε^* satisfying $\varepsilon^* < \frac{\alpha \tau_D}{2\lambda_{\max}(\mathbf{P}) \ln a}$ and $\|\mathbf{L}_{\varepsilon^*}\| < \min_{p \in \mathcal{P}} \frac{\lambda_{\min}(P_p)\lambda_{\min}(\mathbf{P})}{4\lambda_{\max}(P_p)\lambda^2_{\max}(\mathbf{P})}$. It then follows that $\lim_{t \to \infty}(\chi_i(t) - \widehat{\chi}_i(t)) = 0$ for all $i \in \mathcal{V}$.

From the unified observer design, we then have that

$$\widehat{\bar{x}}_i = (T_i^{\mathrm{T}} T_i)^{-1} T_i^{\mathrm{T}} \widehat{\chi}_i = [\widehat{\bar{x}}_{i1}^{\mathrm{T}}, \widehat{\bar{x}}_{i2}^{\mathrm{T}}]^{\mathrm{T}}, \quad i \in \mathcal{V}, \tag{10.13}$$

which will be used in the control design.

10.2.5 Main Results

In this section, we show that the observer architecture introduced in the previous sections provides an asymptotically stable closed-loop system, as presented in Theorem 10.4 below. The observer-based controller is

$$u_i = F_i \widehat{\overline{x}}_{i1} + (\Gamma_i - F_i \Pi_i) \widehat{\overline{x}}_{i2}, \tag{10.14}$$

where Π_i and Γ_i are the solutions of the regulator equations (10.6), and $\widehat{\overline{x}}_{i1}$ and $\widehat{\overline{x}}_{i2}$ can be obtained from (10.8) and (10.13).

Theorem 10.4 *Let Assumptions 10.1–10.3 hold and assume that $\kappa \geq \frac{\alpha + 4 \max\{\theta, 1\} \lambda_{\max}^2(\mathbf{P})}{1 - \alpha}$, where $\alpha \in (0, 1)$, θ, and $\mathbf{P}$ are given by (10.11). Then, there exists $\varepsilon^* \in (0, 1]$ such that, if $\varepsilon \in (0, \varepsilon^*]$, (10.14) guarantees that $\lim_{t \to \infty} e_i(t) = 0$ for all $i \in \mathcal{V}$, for the system (10.1)–(10.4).*

Proof Follows from Lemmas 10.1 and 10.3, and the separation principle.

10.3 Simulations

In this section, we illustrate the theoretical results. Consider a network of three agents. We assume that the adjacency matrix $\bar{A}_{\sigma(t)}$ associated with $\bar{\mathcal{G}}_{\sigma(t)}$ is switching periodically. Denote $\ell = 0, 20, 40 \ldots$. $\bar{A}_1 = \begin{bmatrix} 0 & 0 & 0 & 0 \\ 1 & 0 & 1 & 0 \\ 0 & 1 & 0 & 0 \\ 0 & 0 & 1 & 0 \end{bmatrix}$, when $t \in [\ell, \ell + 6)$, $\bar{A}_2 = \begin{bmatrix} 0 & 0 & 0 & 0 \\ 1 & 0 & 0 & 0 \\ 0 & 1 & 0 & 1 \\ 0 & 0 & 1 & 0 \end{bmatrix}$, when $t \in [\ell + 6, \ell + 12)$, $\bar{A}_3 = \begin{bmatrix} 0 & 0 & 0 & 0 \\ 1 & 0 & 0 & 0 \\ 0 & 1 & 0 & 0 \\ 0 & 0 & 1 & 0 \end{bmatrix}$, when $t \in [\ell + 12, \ell + 18)$, $\bar{A}_4 = \begin{bmatrix} 0 & 0 & 0 & 0 \\ 0 & 0 & 0 & 0 \\ 0 & 0 & 0 & 0 \\ 0 & 0 & 0 & 0 \end{bmatrix}$, when $t \in [\ell + 18, \ell + 20)$.

The dynamics of the agents are described by $\tilde{A}_1 = \begin{bmatrix} 0 & 3 & 0 \\ 0 & 0 & 2 \\ 0 & -1 & 0 \end{bmatrix}$, $\tilde{B}_1 = \begin{bmatrix} 0 \\ 0 \\ 1 \end{bmatrix}$, $C_{s1} = C_{d1} = D_{s1} = \begin{bmatrix} 1 & 1 & 1 \end{bmatrix}$, $\tilde{A}_2 = \begin{bmatrix} 1 & 0 \\ 0 & 0 \end{bmatrix}$, $\tilde{B}_2 = \begin{bmatrix} 1 \\ 1 \end{bmatrix}$, $C_{s2} = \begin{bmatrix} 1 & 0 \end{bmatrix}$, $C_{d2} = \begin{bmatrix} 0 & 1 \end{bmatrix}$, $D_{s2} = \begin{bmatrix} 1 & 1 \end{bmatrix}$, $\tilde{A}_3 = \begin{bmatrix} 0 & 1 \\ -2 & -2 \end{bmatrix}$, $\tilde{B}_3 = \begin{bmatrix} 0 \\ 1 \end{bmatrix}$, $C_{s3} = C_{d3} = D_{s3} = \begin{bmatrix} 1 & 0 \end{bmatrix}$. The dynamics of the individual autonomous exosystems are given by $S_i = 0$, $C_{wi} = D_{wi} = -1$, $i = 1, 2, 3$, and $\omega_1(0) = -2$, $\omega_2(0) = -4$, and $\omega_3(0) = -6$. The dynamics of the group autonomous exosystem are given by $\tilde{A}_0 = \begin{bmatrix} 0 & 1 \\ -1 & 0 \end{bmatrix}$, $C_0 = \begin{bmatrix} 1 & 0 \end{bmatrix}$, $D_0 = -C_0$.

Following the design scheme proposed in Sect. 10.2, for the solutions of regulator equations (10.6), we have that $F_1 = \begin{bmatrix} -1 & -4.5 & -6 \end{bmatrix}$, $\Pi_1 = \begin{bmatrix} 1 & 1.0345 & -0.4138 \\ 0 & 0.1379 & 0.3448 \\ 0 & -0.1724 & 0.0690 \end{bmatrix}$, $\Gamma_1 = \begin{bmatrix} 0 & 0.0690 & 0.1724 \end{bmatrix}$ for agent 1, $F_2 = \begin{bmatrix} -2 & -6 \end{bmatrix}$, $\Pi_2 = \begin{bmatrix} 0 & 0.4 & -0.2 \\ 1 & 0.6 & 0.2 \end{bmatrix}$, $\Gamma_2 = \begin{bmatrix} 0 & -0.2 & 0.6 \end{bmatrix}$ for agent 2, $F_3 = \begin{bmatrix} 0 & -1 \end{bmatrix}$, $\Pi_3 = \begin{bmatrix} 1 & 1 & 0 \\ 0 & 0 & 1 \end{bmatrix}$, $\Gamma_3 = \begin{bmatrix} 2 & 1 & 2 \end{bmatrix}$ for agent 3. We also have $\varepsilon = 0.2$ for (10.8) and $\theta = 0.1$ for (10.11).

Figures 10.3 and 10.4 show, respectively, the state convergence and the error convergence of system (10.1), (10.2), and (10.3) under the observer-based controller (10.14). We see that coordinated output regulation is realized even when there exist

Fig. 10.3 Output convergence of system (10.1), (10.2), and (10.3) under the observer-based controller (10.14)

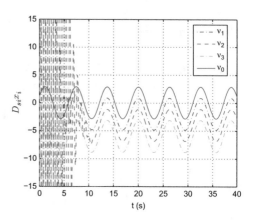

Fig. 10.4 Error convergence of system (10.1), (10.2), and (10.3) under the observer-based controller (10.14)

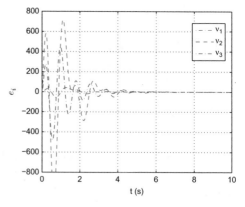

multiple heterogenous dynamics and the information interactions are switching. This agrees with the result in Theorem 10.4.

10.4 Literature

The results in this chapter are based mainly on [9]. Coordinated tracking of networked dynamical systems is also studied in the existing literature. In particular, the authors of [19] generalize coordination of multiple single-integrator systems to the case of multiple linear time-invariant high-order systems. For a network of neutrally stable systems and polynomially unstable systems, the author of [16] proposes a design scheme for achieving synchronization. The case of switching communication topologies is considered in [14] and a so-called consensus-based observer is proposed to guarantee leaderless synchronization of multiple identical linear dynamical systems under a jointly connected communication topology. Similar problems are also considered in [18] and [10], where a frequently connected communication topology is studied in [18] and an assumption on the neutral stability is imposed in [10]. The authors of [6] propose a neighbor-based observer to solve the synchronization problem for general linear time-invariant systems. In addition, the classical Laplacian matrix is generalized in [22] to a so-called interaction matrix and a D-scaling approach is used to stabilize this interaction matrix. Synchronization of multiple heterogeneous linear systems has been investigated under both fixed and switching communication topologies [3, 8, 11, 20]. In [3], a high-gain approach is proposed to dominate the non-identical dynamics of the agents. The cases of frequently connected and jointly connected communication topologies are studied in [17] and [5], respectively, where a slow switching condition and a fast switching condition are presented. Also, the generalizations of coordination of multiple linear dynamical systems to the cooperative output regulation problem are studied in [1, 4, 15, 21].

Acknowledgments ©2015 Elsevier. Reprinted, with permission, from Ziyang Meng, Tao Yang, Dimos V. Dimarogonas, Karl H. Johansson, "Coordinated Output Regulation of Multiple Heterogeneous Linear Systems over Switching Topologies", Automatica, vol. 53, no. 3, 2015, pp. 362–368.

References

1. Z. Ding, Consensus output regulation of a class of heterogeneous nonlinear systems. IEEE Trans. Autom. Control **58**(10), 2648–2653 (2013)
2. B.A. Francis, The linear multivariable regulator problem. SIAM J. Control Optim. **15**(3), 486–505 (1977)
3. H.F. Grip, T. Yang, A. Saberi, et al., Output synchronization for heterogeneous networks of non-introspective agents. Automatica **48**(10), 2444–2453 (2012)

4. H. Kim, H. Shim, J. Seo, Output consensus of heterogeneous uncertain linear multi-agent systems. IEEE Trans. Autom. Control **56**(1), 200–206 (2011)
5. H. Kim, H. Shim, J. Back, et al., Consensus of output-coupled linear multi-agent systems under fast switching network: averaging approach. Automatica **49**(1), 267–272 (2013)
6. Z. Li, Z. Duan, G. Chen, et al., Consensus of multiagent systems and synchronization of complex networks: a unified viewpoint. IEEE Trans. Circuits Syst. I Regular Papers **57**(1), 213–224 (2010)
7. D. Liberzon, A.S. Morse, Basic problems in stability and design of switched systems. IEEE Control Syst. Mag. **19**(5), 59–70 (1999)
8. J. Lunze, Synchronization of heterogeneous agents. IEEE Trans. Autom. Control **57**(11), 2885–2890 (2012)
9. Z. Meng, T. Yang, D.V. Dimarogonas, et al., Coordinated output regulation of heterogeneous linear systems under switching topologies. Automatica **53**(3), 362–368 (2015)
10. W. Ni, D. Cheng, Leader-following consensus of multi-agent systems under fixed and switching topologies. Syst. Control Lett. **59**(3-4), 209–217 (2010)
11. A. Pandey, Output consensus control for heterogeneous multi-agent systems, in *Proceedings of the 52nd IEEE Conf. Decision and Control*, pp. 1502–1507, Florence, Italy, 2015
12. W. Ren, Multi-vehicle consensus with a time-varying reference state. Syst. Control Lett. **56**(7-8), 474–483 (2007)
13. A. Saberi, A.A. Stoorvogel, P. Sannuti, *Control of Linear Systems with Regulation and Input Constraints* (Springer Science and Business Media, London, UK, 2003)
14. L. Scardovi, R. Sepulchre, Synchronization in networks of identical linear systems. Automatica **45**, 546–551 (2009)
15. Y. Su, J. Huang, Cooperative output regulation of linear multi-agent systems. IEEE Trans. Autom. Control **57**(4), 1062–1066 (2012)
16. S.E. Tuna, Conditions for synchronizability in arrays of coupled linear systems. IEEE Trans. Autom. Control **54**(10), 2416–2420 (2009)
17. D. Vengertsev, H. Kim, H. Shim, et al., Consensus of output-coupled linear multi-agent systems under frequently connected network, in *Proceedings of the IEEE Conference on Decision and Control*, pp. 4559–4564, Atlanta, USA, 2010
18. J. Wang, D. Cheng, X. Hu, Consensus of multi-agent linear dynamic systems. Asian J. Control **10**(2), 144–155 (2008)
19. P. Wieland, J.S. Kim, F. Allgöwer, On topology and dynamics of consensus among linear high-order agents. Int. J. Syst. Sci. **42**(10), 1831–1842 (2011)
20. P. Wieland, R. Sepulchre, F. Allgöwer, An internal model principle is necessary and sufficient for linear output synchronization. Automatica **47**(5), 1068–1074 (2011)
21. J. Xiang, W. Wei, Y. Li, Synchronized output regulation of linear networked systems. IEEE Trans. Autom. Control **54**(6), 1336–1341 (2009)
22. T. Yang, S. Roy, Y. Wan, et al., Constructing consensus controllers for networks with identical general linear agents. Int. J. Robust Nonlinear Control **21**(11), 1237–1256 (2011)
23. G. Zhai, B. Hu, K. Yasuda, et al., Piecewise Lyapunov functions for switched systems with average dwell time. Asian J. Control **2**(3), 192–197 (2000)

Part IV
Applications

Chapter 11
Spacecraft Formation Flying

Spacecraft formation flying is an important application of networked dynamical systems. Attitude control of the spacecraft is an indispensable (many applications place strict requirements on it) and difficult (due to the nonlinear dynamics) problem being studied for decades. Attitude is defined as the rotation from one coordinate to a reference coordinate and Fig. 11.1 gives an illustration on the rotation from coordinate (x_b, y_b, z_b) to coordinate (x_a, y_a, z_a). In interferometry applications, it is often essential to control different spacecraft to maintain the same or relative attitudes during and after formation maneuvers, thus bringing much attention to cooperative attitude control problems for multiple spacecraft.

In this chapter, we study the cooperative attitude control problem for underactuated spacecraft. In the first place, we study the leaderless attitude synchronization problem of multiple underactuated spacecraft. We adopt the special parameterizations of attitude proposed by [11] to describe attitude kinematics, which has been shown to be very convenient for control of underactuated axisymmetric spacecraft with two control torques. A partial attitude synchronization controller is proposed. In the second place, full attitude synchronization of multiple underactuated spacecraft is considered, where a discontinuous distributed algorithm is proposed. Simulations are finally given to validate the theoretical results.

11.1   Attitude Kinematics and Dynamics

Since attitude cooperative control problem of multiple underactuated spacecraft is studied in this chapter, we first use the (w, z) parameterization to describe the attitude of underactuated axisymmetric spacecraft. This parameterization is based on a pair of a complex w and a real coordinate z, which is proposed in [11].

© The Author(s), under exclusive license to Springer Nature Switzerland AG 2021
Z. Meng et al., *Modelling, Analysis, and Control of Networked Dynamical Systems*,
Systems & Control: Foundations & Applications,
https://doi.org/10.1007/978-3-030-84682-4_11

Fig. 11.1 An illustration of
attitude

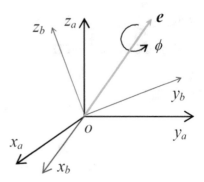

The (w, z) parameterization can be derived from attitude direction cosine matrix $\mathbf{R} = [R_{pq}] \in \mathbb{R}^{3 \times 3}$, $p = 1, 2, 3$, $q = 1, 2, 3$ by using the following relationship (see Lemma 1 of [10]):

$$w = \frac{R_{23}}{1 + R_{33}} - \mathbf{j}\frac{R_{13}}{1 + R_{33}},$$

$$\cos(z) = \frac{1}{2}((1 + |w|^2)\text{trace}(\mathbf{R}) + |w|^2 - 1),$$

where $|w|^2 = w\overline{w}$ denotes the absolute value of $w \in \mathbb{C}$.

Based on the above relationship, we can use (w, z) to describe attitude coordinates from now on. Consider N spacecraft with attitude (w_i, z_i), $\forall i \in V = \{1, 2, \dots, N\}$. Their interactions are described by a time-varying directed graph $\mathcal{G}_{\sigma(t)} = (V, \mathcal{E}_{\sigma(t)})$. The kinematic equation of each spacecraft is described by

$$\dot{w}_i = -\mathbf{j}\omega_{i,3}^* w_i + \frac{\omega_i}{2} + \frac{\overline{\omega_i}}{2}w_i^2, \tag{11.2a}$$

$$\dot{z}_i = \omega_{i,3}^* + \text{Im}(\omega_i \overline{w_i}), \tag{11.2b}$$

where $w_i = w_{i,1} + \mathbf{j}w_{i,2}$ and $\omega_i = \omega_{i,1} + \mathbf{j}\omega_{i,2}$. In this chapter, we focus on angular velocity commands and assume that only angular velocity ω_i can be manipulated. Also, for an axisymmetric spacecraft, $\omega_{i,3}^*$ for all $i \in V$ remains constant for all $t \geq 0$ although the torque input of this axis is zero.

In this case, only two-axis stabilization of pointing is possible for the general case when at least one $\omega_{i,3}^*$, $i \in V$ is nonzero.

In addition, for the special case when $\omega_{i,3}^* \equiv 0$ for all $i \in V$, the three-axis stabilization of pointing is possible, where the kinematic equations become

$$\dot{w}_i = \frac{\omega_i}{2} + \frac{\overline{\omega_i}}{2}w_i^2, \tag{11.3a}$$

$$\dot{z}_i = \text{Im}(\omega_i \overline{w_i}). \tag{11.3b}$$

Also note that every three-dimensional parameterization will cause singularity. Therefore, only almost global attitude control is possible in this chapter and we restrict the discussion on the corresponding kinematic parameters.

11.2 Partial Attitude Synchronization

In this section, we focus on the kinematic (11.2a) and study the partial attitude synchronization problem, where the synchronization manifold $\mathcal{W}$ is defined as $\mathcal{W} = \{(w_1, z_1, \ldots, w_N, z_N) : w_1 = \cdots = w_N\}$.

We first show that partial attitude norm synchronization can be achieved for arbitrary constant $\omega_{i,3}^*$, for all $i \in \mathcal{V}$ under a very mild communication condition, where the norm synchronization manifold $\mathcal{W}_1$ is defined as $\mathcal{W}_1 = \{(w_1, z_1, \ldots, w_N, z_N) : |w_1| = \cdots = |w_N|\}$. We then show that attitude synchronization can be achieved for the special when $\omega_{i,3}^* \equiv 0$ for all $i \in \mathcal{V}$.

The following attitude synchronization algorithm is proposed for the ith spacecraft

$$\omega_i = - \sum_{j \in \mathcal{N}_i(\sigma(t))} a_{ij}(\sigma(t))(w_i - w_j), \qquad (11.4)$$

where $a_{ij}(p) > 0$ is the (i, j)th entry of the adjacency matrix A_p associated with the graph $\mathcal{G}_p$ for all $p \in \mathcal{P}$. It is not hard to show that $a_* \leq a_{ij}(p) \leq a^*$, for all $a_{ij}(p) \neq 0$, all $i, j \in \mathcal{V}$, and all $p \in \mathcal{P}$, where a^* and a_* are defined in Sect. 4.2.

Theorem 11.1 *Suppose $\mathcal{G}_{\sigma(t)}$ is uniformly jointly strongly connected. For multiple underactuated spacecraft kinematic (11.2a), algorithm (11.4) guarantees that partial attitude norm synchronization is achieved with respect to $\mathcal{W}_1$. In particular, it follows that $\lim_{t \to \infty} |w_i(t)| = w^*$ for all $i \in \mathcal{V}$, where w^* is a positive constant.*

Proof Consider the following Lyapunov function candidate

$$V = \max_{i \in \mathcal{V}} V_i,$$

where $V_i = w_i \overline{w_i} = |w_i|^2$ for all $i \in \mathcal{V}$. Note that for all $i \in \mathcal{V}$,

$$\dot{\overline{w_i}} = \overline{\dot{w_i}} = j\omega_{i,3}^* \overline{w_i} + \frac{\overline{\omega_i}}{2} + \frac{\omega_i}{2} \overline{w_i}^2.$$

Therefore, it follows that for all $i \in \mathcal{V}$,

$$\dot{V_i} = \frac{1 + |w_i|^2}{2} \left(\omega_i \overline{w_i} + \overline{\omega_i} w_i \right).$$

Let $\overline{\mathcal{V}}$ be the set containing all the agents that reach the maximum at time t, i.e., $\overline{\mathcal{V}} = \{i \in \mathcal{V}|V_i(t) = V(t)\}$. It follows that the Dini derivative of V can be calculated as

$$D^+V = \max_{i \in \overline{\mathcal{V}}} \dot{V}_i = \max_{i \in \overline{\mathcal{V}}}\left\{\frac{1 + |w_i|^2}{2} \sum_{j \in \mathcal{N}_i(\sigma(t))} a_{ij}(\sigma(t)) \left(w_i\overline{w}_j + \overline{w}_iw_j - 2|w_i|^2\right)\right\}$$

$$\leq \max_{i \in \overline{\mathcal{V}}}\left\{\frac{1 + |w_i|^2}{2} \sum_{j \in \mathcal{N}_i(\sigma(t))} a_{ij}(\sigma(t))(|w_j|^2 - |w_i|^2)\right\} \leq 0,$$

where we have used the fact $w_i\overline{w}_j + \overline{w}_iw_j \leq |w_i|^2 + |w_j|^2$. This implies that $|w_i(t)| \leq V(w(0)) < \infty$, for all $i \in \mathcal{V}$. It follows that $\lim_{t \to \infty} V(t) = V^*$, where V^* is a positive constant. Therefore, we know that for any $\varepsilon > 0$, there exists a $t_1^*(\varepsilon) \geq 0$ such that

$$V_i(t) \leq V^* + \varepsilon, \quad \forall i \in \mathcal{V}, \ \forall t \geq t_1^*.$$

Then, following a similar analysis given in [5], we can show that $V(t^*) < V^* - \varepsilon$ by choosing ε sufficient small for some $t^* \geq t_1^*$. This indicates a contradiction and thus proves the desired result.

We next focus on the special case when the angular velocity of uncontrollable axis $\omega_{i,3}^*$ remains zero for all $i \in \mathcal{V}$. We show that in addition to partial attitude norm synchronization, partial attitude synchronization is also achieved with respect to $\mathcal{W}$.

Theorem 11.2 *Suppose that $\mathcal{G}_{\sigma(t)}$ is uniformly jointly strongly connected. For multiple underactuated spacecraft kinematic (11.2a) with $\omega_{i,3}^* \equiv 0$, for all $i \in \mathcal{V}$, algorithm (11.4) guarantees that partial attitude synchronization is achieved with respect to $\mathcal{W}$. In particular, it follows that $\lim_{t \to \infty}(w_i(t) - w_j(t)) = 0$, for all $i, j \in \mathcal{V}$.*

Proof We show that $\lim_{t \to \infty}(w_i(t) - w_j(t)) = 0$ for all $i, j \in \mathcal{V}$ using a contradiction argument. We still use $V_i = |w_i|^2$ for all $i \in \mathcal{V}$. Based on the result of Theorem 11.1, we know that for any $\varepsilon > 0$, there exists a $t_1^*(\varepsilon) \geq 0$ such that

$$V^* - \varepsilon \leq V_i(t) \leq V^* + \varepsilon, \quad \forall i \in \mathcal{V}, \ \forall t \geq t_1^*,$$

where V^* is a positive constant defined in the proof of Theorem 11.1.

Suppose that there exist $l, k \in \mathcal{V}$ and $t_1 \geq t_1^*$ such that $w_l(t_1) \neq w_k(t_1)$. We next show that there exist $h \in \mathcal{V}$ and $T_1 \geq t_1$ such that $V_h(T_1) < V^* - \varepsilon$, which indicates the contradiction.

Since $\mathcal{G}([t_1, t_1 + T])$ is jointly strongly connected, we can define a time

$$t_2 = \inf_{t \in [t_1, t_1+T]}\{\exists i_1 \in \mathcal{V}\setminus\{l\}|(l, i_1) \in \mathcal{E}_{\sigma(t)}\},$$

and a set

$$\mathcal{V}_1 = \{i_1 \in \mathcal{V}\backslash\{l\} | (l, i_1) \in \mathcal{E}_{\sigma(t_2)}\} \neq \emptyset.$$

Note that for all $i \in \mathcal{V}$, it follows that for all $t \geq t_1$

$$\dot{V}_i = \frac{1 + |w_i|^2}{2} \sum_{j \in \mathcal{N}_i(\sigma(t))} a_{ij}(\sigma(t)) \left(w_i \overline{w_j} + \overline{w_i} w_j - 2|w_i|^2 \right)$$

$$= \sum_{j \in \mathcal{N}_i(\sigma(t))} b_{ij}(w_i, t) \left(-|w_i - w_j|^2 - (|w_i|^2 - |w_j|^2) \right)$$

$$\leq - \sum_{j \in \mathcal{N}_i(\sigma(t))} b_{ij}(w_i, t)|w_i - w_j|^2 + 2b^*(N - 1)\varepsilon,$$

where we define $b_{ij}(w_i, t) = \frac{1+|w_i|^2}{2}a_{ij}(t)$ and easily derive that $b_* \triangleq \frac{1}{2}a_* \leq b_{ij} \leq \frac{1+V(w(0))}{2}a^* \triangleq b^*$ for all $i, j \in \mathcal{V}$. Also, we have used the fact that $\left||w_i(t)|^2 - |w_j(t)|^2\right| = |V_i(t) - V_j(t)| < 2\varepsilon$ for all $i, j \in \mathcal{V}$ and $t \geq t_1$. We next consider two cases:

Case I: There exists a $i_1 \in \mathcal{V}_1$ such that $w_l(t_2) \neq w_{i_1}(t_2)$. It then follows from the definition of t_2 that for all $t \in [t_2, t_2 + \tau_D]$,

$$\dot{V}_l \leq - b_*|w_l - w_{i_1}|^2 + 2b^*(N - 1)\varepsilon.$$

Define a positive constant $\phi^* = |w_l(t_2) - w_{i_1}(t_2)|^2 > 0$ and a function $\phi(t) = |w_\mu(t) - w_\nu(t)|^2$ for any pair $(\mu, \nu) \in \mathcal{V} \times \mathcal{V}$. By noting the fact that $|w_\mu(t) - w_\nu(t)|^2 \leq 2|w_\mu(t)|^2 + 2|w_\nu(t)|^2 = 2(V_\mu(t) + V_\nu(t))$ for all $t \geq 0$, we next bound the trajectory of $\phi(t)$ by analyzing the dynamics of V_i, $i \in \mathcal{V}$. Since $b_{ij} \leq b^*$, for all $i, j \in \mathcal{V}$ and $V_i(t) \leq V^* + \varepsilon$, $\forall i \in \mathcal{V}$, it follows that

$$\dot{V}_i \leq 2b^*(N - 1)(V^* + 2\varepsilon), \ \forall i \in \mathcal{V}, \ \forall t \geq t_1.$$

It thus follows that $\phi(t) \geq \frac{\phi^*}{2}$ for all $t \in [t_2, t_2 + \frac{\phi^*}{16b^*(N-1)(V^*+2\varepsilon)}]$.
Define $\delta_1 = \min\{\tau_D, \frac{\phi^*}{16b^*(N-1)(V^*+2\varepsilon)}\}$. It then follows that

$$V_l(t_2 + \delta_1) \leq V_l(t_2) - \frac{b_*\phi^*\delta_1}{2} + 2b^*(N - 1)\delta_1\varepsilon$$

$$\leq V^* - \frac{b_*\phi^*\delta_1}{2} + (2b^*(N - 1)\delta_1 + 1)\varepsilon.$$

We then know that $V_l(t_2 + \tau_D) < V^* - \varepsilon$ by choosing ε sufficiently small. This proves the desired result.

Case II: For all $i_1 \in \mathcal{V}_1$, $w_{i_1}(t_2) = w_l(t_2)$. Then, based on the definitions of t_2 and $\mathcal{V}_1$, we know that the dynamics $\dot{w}_l = \frac{\omega_l}{2} + \frac{\overline{\omega}_l}{2} w_l^2$ is reduced to

$$\dot{w}_l = 0, \ \forall t \in [t_1, t_2].$$

Therefore, we know that $w_l(t_2) = w_l(t_1)$ and thus $w_{i_1}(t_2) = w_l(t_1)$, for all $i_1 \in \mathcal{V}_1$.

Next, we define the set $\hat{\mathcal{V}}_1 = \{l\} \bigcup \mathcal{V}_1$. Since $\mathcal{G}([t_1 + T, t_1 + 2T])$ is jointly strongly connected, we can define a time

$$t_3 = \inf_{t \in [t_2, t_1 + 2T]} \{\exists i_2 \in \mathcal{V} \backslash \hat{\mathcal{V}}_1, i_1 \in \hat{\mathcal{V}}_1 | (i_1, i_2) \in \mathcal{E}_{\sigma(t)}\},$$

and a set

$$\mathcal{V}_2 = \{i_2 \in \mathcal{V} \backslash \hat{\mathcal{V}}_1 | (i_1, i_2) \in \mathcal{E}_{\sigma(t_3)}\}.$$

Following the above analysis, we can show that $V_{i_1}(t_3 + \delta_2) < V^* - \varepsilon$ for some $i_1 \in \mathcal{V}_1$ when $w_{i_2}(t_3) \neq w_{i_1}(t_3)$, for some $i_2 \in \mathcal{V}_2$. Otherwise, for the case that $w_{i_2}(t_3) = w_{i_1}(t_3)$ for all $i_2 \in \mathcal{V}_2$, we know that $w_{i_1}(t)$ remains unchanged for all $i_1 \in \mathcal{V}_1$ during $t \in [t_2, t_3]$ based on the definitions of t_3 and $\mathcal{V}_2$. It then follows that $w_{i_1}(t_3) = w_{i_1}(t_2) = w_l(t_1)$ for all $i_1 \in \mathcal{V}_1$ and thus $w_{i_2}(t_3) = w_l(t_1)$ for all $i_2 \in \mathcal{V}_2$.

Then, we define $\hat{\mathcal{V}}_2 = \hat{\mathcal{V}}_1 \bigcup \mathcal{V}_2$ and repeat the above analysis. At worst case, we finally have $\hat{\mathcal{V}}_{N-2} = \mathcal{V} \backslash \{k\}$ and the time

$$t_n = \inf_{t \in [t_{N-1}, t_1 + (N-1)T]} \{\exists i_{N-2} \in \hat{\mathcal{V}}_{N-2} | (i_{N-2}, k) \in \mathcal{E}_{\sigma(t)}\}.$$

By noting that $w_{i_{N-2}}(t_N) = w_l(t_1)$, for all $i_{N-2} \in \hat{\mathcal{V}}_{N-2}$ and the earliest assumption of $w_l(t_1) \neq w_k(t_1)$, we know that $w_k(t_N) \neq w_{i_{N-2}}(t_N)$, for some $i_{N-2} \in \hat{\mathcal{V}}_{N-2}$. Therefore, we can show that $V_{i_{N-2}}(t_N + \tau_D) < V^* - \varepsilon$ by choosing ε sufficient small. This proves the desired result.

11.3   Full Attitude Synchronization

In this section, we focus on the kinematic (11.3) and study the full attitude synchronization problem, where the synchronization manifold $\mathcal{P}$ is defined as $\mathcal{P} = \{(w_1, z_1, \ldots, w_N, z_N) : z_1 = \cdots = z_N, w_1 = \cdots = w_N\}$.

The following attitude synchronization algorithm is proposed for the ith spacecraft

$$\omega_i = -\gamma w_i - \mathbf{j}\frac{\sum_{j\in\mathcal{N}_i(\sigma(t))} a_{ij}(\sigma(t))(z_i - z_j)}{\overline{w_i}}, \tag{11.5}$$

where $a_{ij}(\sigma(t)) > 0$ is the weight of edge (i, j) for $i, j \in V$ at t.

Theorem 11.3 *Suppose that $\mathcal{G}_{\sigma(t)}$ is uniformly jointly strongly connected. Also assume that $w_i(0) \neq 0$, for all $i \in V$. For multiple underactuated spacecraft (11.3), algorithm (11.5) guarantees that full attitude synchronization is achieved with respect to $\mathcal{P}$ provided that γ is sufficiently small. In particular, we have that:*

- *$w_i(t) \neq 0$, for all $i \in V$ and for all $t \geq 0$.*
- *$\lim_{t\to\infty} w_i(t) = 0$ for all $i \in V$.*
- *$\lim_{t\to\infty} z_i(t) = z^*$ for all $i \in V$, where z^* is a positive constant.*
- *The control input ω_i is bounded for all $i \in V$.*

Proof It is clear that the closed-loop system (11.3) with distributed attitude control algorithm (11.5) can be written as

$$\dot{w}_i = -\frac{\gamma}{2}(1 + |w_i|^2)w_i - \mathbf{j}\frac{1}{2}\sum_{j\in\mathcal{N}_i(\sigma(t))} a_{ij}(\sigma(t))(z_i - z_j)\left(\frac{1}{\overline{w_i}} - w_i\right), \tag{11.6a}$$

$$\dot{z}_i = -\sum_{j\in\mathcal{N}_i(\sigma(t))} a_{ij}(\sigma(t))(z_i - z_j), \quad i \in V. \tag{11.6b}$$

Note that the above equations hold for $(\mathbb{C}\backslash\{0\}) \times \mathbb{R} \times \cdots \times (\mathbb{C}\backslash\{0\}) \times \mathbb{R}$. We first show that $w_i(t) \neq 0$, for all $i \in V$ and for all $t \geq 0$ given $w_i(0) \neq 0$, for all $i \in V$.

Consider the Lyapunov function candidate $V_i = |w_i|^2$ for all $i \in V$. The derivative of V_i along (11.6a) can be calculated as

$$\dot{V}_i = 2\text{Re}(\dot{w}_i\overline{w_i}) = -\gamma(1 + |w_i|^2)|w_i|^2.$$

It thus follows that

$$|w_i| = \sqrt{\frac{1}{c_i e^{\gamma t} - 1}},$$

where $c_i = \frac{1+|w_i(0)|^2}{|w_i(0)|^2}$ is a bounded constant for all $i \in V$. Therefore, $w_i(t) \neq 0$ for all $t \geq 0$ and we know that $\lim_{t\to\infty} w_i(t) = 0$ for all $i \in V$.

For system (11.6b), we can show that $\lim_{t\to\infty} z_i(t) = z^*$, where z^* is a positive constant, for all $i \in V$ for all $i \in V$ by choosing Lyapunov function $V_p = \max_{i\in V} z_i^2$ and following a similar analysis on the proof of Theorem 11.1.

We next show that the control input ω_i is bounded for all $i \in V$. Consider the following Lyapunov function:

$$V_q = \max_{\{i,j\} \in \mathcal{V} \times \mathcal{V}} V_{ij},$$

where $V_{ij}(z) = (z_i(t) - z_j(t))^2$ for all $\{i, j\} \in \mathcal{V} \times \mathcal{V}$ and $z = [z_1, z_2, \dots, z_N]^{\mathrm{T}}$.

Let $\mathcal{V}^1 \times \mathcal{V}^2$ be the set containing all the agent pairs that reach the maximum at time t, i.e., $\mathcal{V}^1(t) \times \mathcal{V}^2(t) = \{\{i, j\} \in \mathcal{V} \times \mathcal{V} | V_{ij}(t) = V_q(t)\}$. The derivative of V_q can be calculated as

$$
\begin{aligned}
D^+ V_q &= \max_{\{i,j\} \in \mathcal{V}^1 \times \mathcal{V}^2} (z_i - z_j)^{\mathrm{T}} \sum_{k \in \mathcal{N}_i(\sigma(t))} a_{ik}(\sigma(t))(z_i - z_k) \\
&\quad - (z_i - z_j)^{\mathrm{T}} \sum_{k \in \mathcal{N}_j(\sigma(t))} a_{jk}(\sigma(t))(z_j - z_k) \\
&\leq - \max_{\substack{\{i,j\} \in \mathcal{V}^1 \times \mathcal{V}^2 \\ j \in \mathcal{N}_i(\sigma(t))}} \sum a_{ik}(\sigma(t))((z_i - z_j)^2 - (z_j - z_k)^2) \\
&\quad - \sum_{k \in \mathcal{N}_j(\sigma(t))} a_{jk}(\sigma(t))((z_j - z_i)^2 - (z_i - z_k)^2) \\
&\leq - \max_{\{i,j\} \in \mathcal{V}^1 \times \mathcal{V}^2} \left\{ \sum_{j \in \mathcal{N}_i(\sigma(t))} a_{ik}(\sigma(t))(V_{ij} - V_{jk}) + \sum_{k \in \mathcal{N}_j(\sigma(t))} a_{jk}(\sigma(t))(V_{ji} - V_{ik}) \right\} \\
&\leq 0,
\end{aligned}
$$

where we have used the fact that $V_{ij} = V_{ji}$. Using comparison lemma [4], we know that $V_q(z) \leq V_q(0)$.

Define $\overline{T} = T + 2\tau_D$. Consider time interval $[0, \bar{N}\overline{T}]$, where $\bar{N} = \frac{N(N-1)}{2}$. It follows that for all $t \in [0, \bar{N}\overline{T}]$,

$$V_{ij}(z) \leq V_q(0), \quad \forall \{i, j\} \in \mathcal{V} \times \mathcal{V}.$$

We first consider any agent i_1 and focus on the time interval $[0, \overline{T}]$. It follows from uniformly jointly strongly connected assumption that i_1 is the root and there exists a time such that there is an agent i_2 being the neighbor of agent i_1 during $t \in [T_1, T_1 + \tau_D] \subset [0, \overline{T}]$. Then for all $t \in [T_1, T_1 + \tau_D]$, it follows that

$$
\begin{aligned}
\dot{V}_{i_1 i_2}(t) &= - (z_{i_1} - z_{i_2})^{\mathrm{T}} \sum_{k \in \mathcal{N}_{i_1}(\sigma(t))} a_{i_1 k}(\sigma(t))(z_{i_1} - z_k) \\
&\quad + (z_{i_1} - z_{i_2})^{\mathrm{T}} \sum_{k \in \mathcal{N}_{i_2}(\sigma(t))} a_{i_2 k}(\sigma(t))(z_{i_2} - z_k) \\
&\leq - \sum_{k \in \mathcal{N}_{i_1}(\sigma(t))} a_{i_1 k}(\sigma(t))(V_{i_1 i_2} - V_{i_2 k}) - 2a_{i_2 i_1}(t) V_{i_1 i_2}
\end{aligned}
$$

$$- \sum_{k \in \mathcal{N}_{i_2}(\sigma(t)) \setminus \{i_1\}} a_{i_2 k}(\sigma(t))(V_{i_1 i_2} - V_{i_1 k})$$

$$\leq - (N-1)a^*(V_{i_1 i_2} - V_q(0)) - (N-2)a^*(V_{i_1 i_2} - V_q(0)) - a_* V_{i_1 i_2}$$

$$\leq - \alpha(V_{i_1 i_2}(t) - \frac{2(N-1)a^*}{\alpha} V_q(0)),$$

where $\alpha = 2(N-1)a^* + a_*$. It thus follows that

$$V_{i_1 i_2}(T_1 + \tau_D) \leq (1 - e^{-\alpha \tau_D}) \frac{2(N-1)a^*}{\alpha} V_q(0) + e^{-\alpha \tau_D} V_{i_1 i_2}(T_1)$$

$$\leq ((1 - e^{-\alpha \tau_D}) \frac{2(N-1)a^*}{\alpha} + e^{-\alpha \tau_D}) V_q(0) = \beta_1 V_q(0).$$

Note that $\beta_1 = (1 - e^{-\alpha \tau_D}) \frac{2(N-1)a^*}{\alpha} + e^{-\alpha \tau_D} < 1$.

It then follows that for all $t \in [T_2 + \tau_D, \bar{N}\bar{T}]$,

$$\dot{V}_{i_1 i_2} \leq - \sum_{k \in \mathcal{N}_{i_1}(\sigma(t))} a_{i_1 k}(\sigma(t))(V_{i_1 i_2} - V_{i_2 k}) - \sum_{k \in \mathcal{N}_{i_2}(\sigma(t))} a_{i_2 k}(\sigma(t))(V_{i_1 i_2} - V_{i_1 k})$$

$$\leq - 2(N-1)a^*(V_{i_1 i_2} - V_q(0)).$$

This implies that for all $t \in [T_2 + \tau_D, \bar{N}\bar{T}]$,

$$V_{i_1 i_2}(t) \leq \left((1 - e^{-\alpha_2 \bar{N}\bar{T}}) + e^{-\alpha_2 \bar{N}\bar{T}} \beta_1\right) V_q(0),$$

where $\alpha_2 = 2(N-1)a^*$. Note that $1 - (1 - \beta_1)e^{-\alpha_2 \bar{N}\bar{T}} < 1$. Therefore, we know that for all $t \in [T_1 + \tau_D, \bar{N}\bar{T}]$,

$$V_{i_1 i_2}(t) \leq (1 - (1 - e^{-\alpha \tau_D}) \frac{2a_*}{\alpha} e^{-\alpha_2 \bar{N}\bar{T}}) V_q(0).$$

We next focus on the time interval $[\bar{T}, 2\bar{T}]$ and consider $\mathcal{V}_1 = \{i_1, i_2\}$ together. We know from strongly connected assumption that there exists an edge from $j \in \mathcal{V}_1$ to i_3 during $[T_2, T_2 + \tau_D] \subset [\bar{T}, 2\bar{T}]$, where $T_2 \in [\bar{T}, \bar{T} + T]$. We next bound $V_{i_1 i_3}$. We consider two cases:

Case I: $j = i_1$. Following a similar analysis on $V_{i_1 i_2}$, we know that for all $t \in [T_2 + \tau_D, \bar{N}\bar{T}]$,

$$V_{i_1 i_3}(t) \leq (1 - (1 - e^{-\alpha \tau_D}) \frac{2a_*}{\alpha} e^{-\alpha_2 \bar{N}\bar{T}}) V_q(0).$$

Case II: $j \in \mathcal{V}_1 \setminus \{i_1\}$. It then follows that for all $t \in [T_2, T_2 + \tau_D]$,

$$
\begin{aligned}
\dot{V}_{i_1 i_3}(t) \leq &- \sum_{k \in \mathcal{N}_{i_1}(\sigma(t))} a_{i_1 k}(\sigma(t))(V_{i_1 i_3} - V_{i_3 k}) + \sum_{k \in \mathcal{N}_{i_3}(\sigma(t))} a_{i_3 k}(\sigma(t))(V_{i_1 i_3} - V_{i_1 k}) \\
\leq &- (N-1)a^*(V_{i_1 i_3} - V_q(0)) - (N-2)a^*(V_{i_1 i_3} - V_q(0)) \\
&- a_{i_3 j}(V_{i_1 i_3} - V_{i_1 j}) \\
\leq &- 2(N-1)a^*(V_{i_1 i_3} - V_q(0)) - a_{i_3 j}(V_{i_1 i_3} - V_{i_1 j}).
\end{aligned}
$$

This also involves two subcases:

Case II.(a): $V_{i_1 i_3}(t) > V_{i_1 j}(t)$ for all $t \in [T_2, T_2 + \tau_D]$. It then follows that

$$
\begin{aligned}
\dot{V}_{i_1 i_3}(t) \leq &- 2(N-1)a^*(V_{i_1 i_3} - V_q(0)) - a_*(V_{i_1 i_3} - \bar{\beta} V_q(0)) \\
\leq &- \alpha \left(V_{i_1 i_3} - \frac{2(N-1)a^* + a_* \bar{\beta}}{\alpha} V_q(0) \right),
\end{aligned}
$$

where $\bar{\beta} = (1 - (1 - e^{-\alpha \tau_D}) \frac{a_*}{\alpha} e^{-\alpha_2 \bar{N} \bar{T}})$. This shows that

$$
V_{i_1 i_3}(T_2 + \tau_D) \leq (1 - e^{-\alpha \tau_D}) \frac{2(N-1)a^* + a_* \bar{\beta}}{\alpha} V_q(0) + e^{-\alpha \tau_D} V_q(0).
$$

It follows that for all $t \in [T_2 + \tau_D, \bar{N} \bar{T}]$,

$$
V_{i_1 i_3}(t) \leq (1 - (1 - e^{-\alpha \tau_D}) \frac{(1 - \bar{\beta})a_*}{\alpha} e^{-\alpha_2 \bar{N} \bar{T}}) V_q(0) = (1 - \beta^2) V_q(0),
$$

where $\beta = (1 - e^{-\alpha \tau_D}) \frac{a_*}{\alpha} e^{-\alpha_2 \bar{N} \bar{T}}$.

Case II.(b):   There exists a time $t^* \in [T_2, T_2 + \tau_D]$ such that $V_{i_1 i_3}(t^*) \leq V_{i_1 j}(t^*)$. It then follows that for all $t \in [t^*, \bar{N} \bar{T}]$,

$$
V_{i_1 i_3}(t) \leq \left((1 - e^{-\alpha_2 \bar{N} \bar{T}}) + e^{-\alpha_2 \bar{N} \bar{T}} \bar{\beta} \right) V_q(0).
$$

Combining the above analysis, we know that for all $t \in [T_2 + \tau_D, \bar{N} \bar{T}]$,

$$
V_{i_1 i_3}(t) \leq (1 - \beta^2) V_q(0).
$$

Next, we consider $\mathcal{V}_2 = \{i_1, i_2, i_3\}$ together. We have shown that for all $t \in [T_2 + \tau_D, \bar{N} \bar{T}]$,

$$
V_{i_1 k}(t) \leq (1 - \beta^2) V_q(0),
$$

for all $k \in V_2 \setminus \{i_1\}$. Continuing the above analysis, we can show that for all $t \in [T_{N-1} + \tau_D, \bar{N}\bar{T}]$,

$$V_{i_1 k}(t) \leq (1 - \beta^{N-1}) V_q(0), \tag{11.7}$$

for all $k \in V \setminus \{i_1\}$, where $T_{N-1} \in [(N-2)\overline{T}, (N-2)\overline{T} + T]$.

Then, we consider i_2 as the root. Note that we have shown that $V_{i_2 i_1}$ satisfies (11.7). Therefore, we can similarly show that for all $t \in [T_{2N-3} + \tau_D, \bar{N}\bar{T}]$,

$$V_{i_2 k}(t) \leq (1 - \beta^{2N-3}) V_q(0),$$

for all $k \in V \setminus \{i_2\}$, where $T_{2N-3} \in [(2N-4)\overline{T}, (2N-4)\overline{T} + T]$.

Similar to the above analysis, we can show that $t \in [T_{N(N-1)/2} + \tau_D, \bar{N}\bar{T}]$,

$$V_{ij}(t) \leq (1 - \beta^{N(N-1)/2}) V_q(0),$$

for all $i, j \in V$, where $T_{N(N-1)/2} \in [(N(N-1)/2-1)\overline{T}, (N(N-1)/2-1)\overline{T}+T]$. Finally, we know that

$$V_q(\bar{N}\bar{T}) - V_q(0) \leq -\beta^{N(N-1)/2} V_q(0).$$

Then, let ψ be the smallest positive integer satisfying $t \leq \psi \bar{N}\bar{T}$. It then follows that

$$V_q(t) \leq (1 - \beta^{N(N-1)/2})^{\psi-1} V_q(0) \leq \frac{1}{1-\beta}(1 - \beta^{N(N-1)/2})^{\frac{t}{N\bar{T}}} V_q(0)$$

$$= \frac{1}{1-\beta} e^{-\beta^* t} V_q(0),$$

where $\beta^* = \frac{1}{\bar{N}\bar{T}} \ln \frac{1}{1-\beta^{N(N-1)/2}}$. This shows that

$$V_q(t) \leq \mu e^{-\beta^* t} V_q(0),$$

where $\mu = \frac{2}{1-\beta}$.

We can now use the above fact to show that ω_i is bounded, for all $i \in V$. It is trivial to see that for all $i \in V$

$$|\omega_i(t)| \leq \gamma |w_i(t)| + \frac{(N-1)a^* \max_{\{i,j\} \in V \times V} |z_i(t) - z_j(t)|}{|w_i(t)|}$$

$$\leq \gamma |w_i(0)| + \sqrt{\mu}(N-1)a^* \sqrt{V_q(0)} e^{-0.5\beta^* t} \sqrt{c_i e^{\gamma t} - 1}$$

$$< \gamma |w_i(0)| + \sqrt{\mu}(N-1)a^* \sqrt{c_i} \sqrt{V_q(0)} e^{-0.5(\beta^* - \gamma)t}.$$

Therefore,

$$|\omega_i| \leq \gamma|w_i(0)| + \sqrt{\mu}(N-1)a^*\sqrt{c_i}\sqrt{V_q(0)},$$

if we choose that $\gamma < \beta^*$. To this end, the desired results have all been proven.

11.4 Simulations

In this section, we present two examples to illustrate the theoretical results. In particular, we consider that there are four spacecraft ($N = 4$) in the group. The weight a_{ij} is chosen to be 1 when $(i, j) \in \mathcal{E}$. The communication graph $\mathcal{G}$ switches between $\mathcal{G}_1$ (Fig. 11.2) and $\mathcal{G}_2$ (Fig. 11.3) at time instants $t_\ell = \ell, \ell = 0, 1, \ldots$.

11.4.1 Partial Attitude Synchronization

We first consider multiple underactuated spacecraft kinematic (11.2a) under algorithm (11.4) and the angular velocities $\omega_{i,3}^*$ are zeros for all $i \in V$. Figure 11.4 shows the trajectories of $w_{i,1}$ and $w_{i,2}$ for all $i = 1, 2, 3, 4$. We see that attitude synchronization is achieved, while the final attitudes of all the spacecraft converge to a nonzero constant. This agrees with the results of Theorem 11.2.

Fig. 11.2 The communication graph $\mathcal{G}_1$

Fig. 11.3 The communication graph $\mathcal{G}_2$

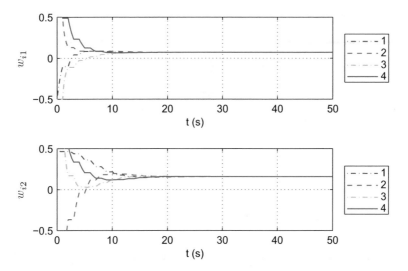

Fig. 11.4 Attitude trajectories of system (11.2a) under algorithm (11.4)

11.4.2   Full Attitude Synchronization

We first consider multiple underactuated spacecraft (11.3) under algorithm (11.5). The control gain γ is chosen as $\gamma = 0.1$. Figures 11.5 and 11.6 show, respectively, the trajectories of $w_{i,1}$, $w_{i,2}$, and z_i and the control inputs $\omega_{i,1}$ and $\omega_{i,2}$, for all $i = 1, 2, 3, 4$. We see that attitude synchronization is achieved, while the control inputs are bounded. This agrees with the results of Theorem 11.3.

11.5   Literature

The results in this chapter are based mainly on [6]. For further results on cooperative attitude control of multiple spacecraft, see [1, 2, 7–9, 12]. In particular, attitude synchronization problem of a group of rotating and translating rigid bodies is studied in [7], where a ring communication topology is considered. Attitude direction cosine matrix is used in [9] to construct an attitude synchronization algorithm, while the extension to the case of directed switching communication topologies is also given. The authors of [12] propose a cooperative attitude tracking protocol such that the follower spacecraft track a time-varying leader spacecraft using relative attitude and relative angular velocity information. A passivity-based group orientation approach is introduced in [1] to solve distributed attitude alignment problem, where the inertial frame information is not assumed to be available to the spacecraft. Similar problem is also considered in [8], where a standing assumption is that the states of the leader spacecraft are only available to a subset of follower spacecraft and the

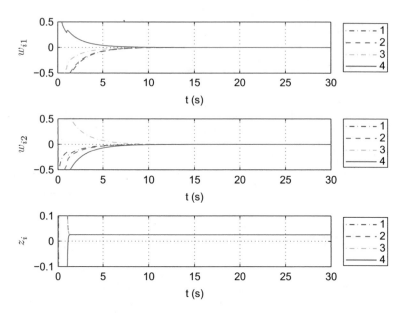

Fig. 11.5 Attitude trajectories of system (11.3) under algorithm (11.5)

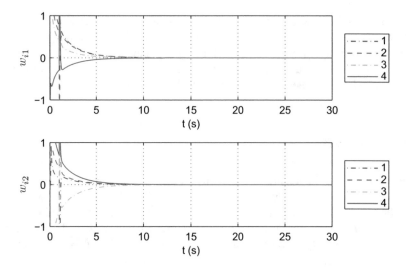

Fig. 11.6 Trajectories of control inputs of system (11.3) under algorithm (11.5)

follower spacecraft only have local information exchange. In addition, the attitude containment problem is considered in [2] and the influence of communication delays between different spacecraft is studied in [3].

Acknowledgments ©2017 IEEE. Reprinted, with permission, from Ziyang Meng, Dimos V. Dimarogonas, Karl H. Johansson, "Attitude synchronization of multiple underactuated axisymmetric spacecraft", IEEE Transactions on Control of Network Systems, vol. 4, no. 4, pp. 816–825, 2017.

References

1. H. Bai, M. Arcak, J.T. Wen, Rigid body attitude coordination without inertial frame information. Automatica **44**(12), 3170–3175 (2008)
2. D.V. Dimarogonas, P. Tsiotras, K.J. Kyriakopoulos, Leader–follower cooperative attitude control of multiple rigid bodies. Syst. Control Lett. **58**(6), 429–435 (2009)
3. E. Jin, X. Jiang, Z. Sun, Robust decentralized attitude coordination control of spacecraft formation. Syst. Control Lett. **57**(7), 567–577 (2008)
4. H.K. Khalil, *Nonlinear Systems, Third Edition* (Prentice-Hall, USA, 2002)
5. Z. Meng, T. Yang, G. Shi, et al., Set target aggregation of multiple mechanical systems, in *Proceedings of the IEEE Conference on Decision and Control*, pp. 6830–6835, Los Angeles, USA, 2014
6. Z. Meng, D.V. Dimarogonas, K.H. Johansson, Attitude coordinated control of multiple underactuated axisymmetric spacecraft. IEEE Trans. Control Netw. Syst. **4**(4), 816–825 (2017)
7. S. Nair, N.E. Leonard, Stable synchronization of mechanical system networks. Am. Inst. Math. Sci. **47**(2), 661–683 (2008)
8. W. Ren, Distributed cooperative attitude synchronization and tracking for multiple rigid bodies. IEEE Trans. Control Syst. Tech. **18**(2), 383–392 (2010)
9. A. Sarlette, R. Sepulchre, N.E. Leonard, Autonomous rigid body attitude synchronization. Automatica **45**(2), 572–577 (2009)
10. P. Tsiotras, J. Luo, Control of underactuated spacecraft with bounded inputs. Automatica **36**(8), 1153–1169 (2000)
11. P. Tsiotras, M. Corless, J.M. Longuski, A novel approach to the attitude control of axisymmetric spacecraft. Automatica **31**(8), 1099–1112 (1995)
12. M.C. VanDyake, C.D. Hall, Decentralized coordinated attitude control of a formation of spacecraft. J. Guidance Control Dyn. **29**(5), 1101–1109 (2006)

Chapter 12
Multi-robot Rendezvous

The study on multi-robot systems attracts much attention recently due to its potential applications in environment monitoring, search and rescue, and entertainment. See Fig. 12.1 for an illustration of a group of robotics. In this chapter, we study rendezvous or the so-called cooperative set aggregation problem for a group of robotics. In particular, each robot observes a convex set as its local target and the objective of the group is to reach a generalized coordinate agreement toward these target sets. We first consider the case when the communication graphs are fixed. A distributed control law is proposed based on each system's own target sensing and information exchange with neighbors. With necessary connectivity, the generalized coordinates of multiple robotic systems are shown to achieve agreement in the intersection of all local target sets, while generalized coordinate derivatives are driven to zero. Moreover, when communication graphs are allowed to be switching, we propose a model-dependent control algorithm and show that cooperative set aggregation is achieved when joint connectivity is guaranteed and the intersection of local target sets is nonempty. Simulations are given to validate the theoretical results and some discussions are provided on the situation when target sets are non-convex.

12.1 Problem Formulation

Consider a network with N agents. As presented in Sect. 3.4.2, the dynamics of agent $i \in \mathcal{V}$ is described by the Lagrangian equation (3.5). Also note that the dynamics of a Lagrangian system i satisfies the following properties [17]:

1. $M_i(q_i)$ is positive definite and is bounded for any $q_i \in \mathbb{R}^n$. More specifically, there exist positive constants $k_{\overline{M}}$ and $k_{\underline{M}}$ such that $k_{\underline{M}} I_n \leq M_i(q_i) \leq k_{\overline{M}} I_n$.
2. $\dot{M}_i(q_i) - 2C_i(q_i, \dot{q}_i)$ is skew symmetric.

© The Author(s), under exclusive license to Springer Nature Switzerland AG 2021   141
Z. Meng et al., *Modelling, Analysis, and Control of Networked Dynamical Systems*,
Systems & Control: Foundations & Applications,
https://doi.org/10.1007/978-3-030-84682-4_12

Fig. 12.1 Robotic network

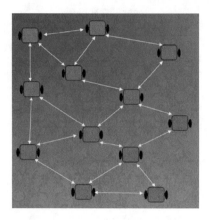

Fig. 12.2 Targeted
aggregation problem

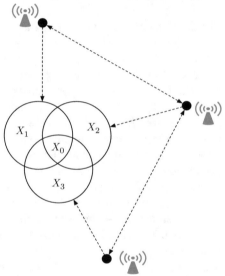

3. $C_i(q_i, \dot{q}_i)$ is bounded with respect to q_i and linearly bounded with respect to $\dot{q}_i$.
   More specifically, there is positive constant k_C such that $\|C_i(q_i, \dot{q}_i)\| \leq k_C \|\dot{q}_i\|$.

We consider the cooperative set aggregation problem for a group of robotic
systems. Each agent $i \in \mathcal{V}$ observes its own target set $\mathcal{X}_i$. The objective is to
ensure that the generalized coordinate derivatives of all the agents converge to
zero and their generalized coordinates achieve agreement, while the destination
of each agent is constrained by its target set. At each time, we assume that each
agent observes the boundary points of its target set and obtains the relative distance
information between the target set and itself. Also, the state information of each
agent is exchanged by equipping each agent with simple and cheap communication
unit. The sketch map of the cooperative set aggregation problem is presented in
Fig. 12.2.

We impose an assumption on set $\mathcal{X}_i, i \in \mathcal{V}$.

Assumption 12.1 $\mathcal{X}_1, \mathcal{X}_2, \ldots, \mathcal{X}_N$ *are compact convex sets.*

Note that Assumption 12.1 is very general and has been extensively used in the existing literature. We next introduce the definition on cooperative set aggregation, where all the agents not only reach an agreement, but also converge to the intersection of all $\mathcal{X}_i, i \in \mathcal{V}$, i.e., $\mathcal{X}_0 = \bigcap_i \mathcal{X}_i$.

Definition 12.1 Multi-robot system (3.5) with a given control law τ_i, for all $i \in \mathcal{V}$, achieves cooperative set aggregation if for all $q_i(0) \in \mathbb{R}^n$, $\dot{q}_i(0) \in \mathbb{R}^n$, $i \in \mathcal{V}$:

1. $\lim_{t \to \infty} d(q_i(t), \mathcal{X}_0) = 0$, $\forall i \in \mathcal{V}$
2. $\lim_{t \to \infty}(q_i(t) - q_j(t)) = 0$, $\quad \forall i, j \in \mathcal{V}$
3. $\lim_{t \to \infty} \dot{q}_i(t) = 0$, $\quad \forall i \in \mathcal{V}$

12.2 Fixed Graphs

Let an undirected graph $\mathcal{G} = (\mathcal{V}, \mathcal{E})$ define the communication topology. Moreover, recall that j is a neighbor of i when $\{j, i\} \in \mathcal{E}$, and $\mathcal{N}_i$ represents the set of agent i's neighbors.

The model-independent control law is proposed for all $i \in \mathcal{V}$:

$$\tau_i = -k_i \dot{q}_i - \alpha_i(q_i - P_{\mathcal{X}_i}(q_i)) - \beta \sum_{j \in \mathcal{N}_i} a_{ij}(q_i - q_j), \tag{12.1}$$

where $k_i > 0$ denotes generalized coordinate derivative damping, $q_i - P_{\mathcal{X}_i}(q_i)$ denotes the relative distance between q_i to the set $\mathcal{X}_i$, $\alpha_i > 0$ denotes the gain for the target set projection control, $\beta > 0$ denotes the gain for the cooperative control, and $a_{ij} > 0$ is the (i, j)th entry of the adjacency matrix A associated with the graph $\mathcal{G}$, which marks the strength of the information flow between i and j.

We next present the main results of this section.

Theorem 12.2 *Suppose that Assumption 12.1 holds and the fixed graph $\mathcal{G}$ is connected. The multi-robotic system (3.5) with (12.1) achieves cooperative set aggregation in the sense of Definition 12.1 if and only if $\mathcal{X}_0$ is nonempty.*

Proof (Sufficiency.) Note that the closed-loop system (3.5) with (12.1) can be written by

$$\dot{q}_i = \dot{q}_i, \tag{12.2a}$$

$$\ddot{q}_i = M_i^{-1}(q_i) \left(C_i(q_i, \dot{q}_i) \dot{q}_i - k_i \dot{q}_i - \alpha_i(q_i - P_{\mathcal{X}_i}(q_i)) \right.$$

$$-\beta \sum_{j \in \mathcal{N}_i} a_{ij}(q_i - q_j)\bigg), \quad i \in \mathcal{V}. \tag{12.2b}$$

Choose state variable as $x = [q^{\mathrm{T}}, \dot{q}^{\mathrm{T}}]^{\mathrm{T}}$, where $x = [q_1^{\mathrm{T}}, q_2^{\mathrm{T}}, \ldots, q_N^{\mathrm{T}}, \dot{q}_1^{\mathrm{T}}, \dot{q}_2^{\mathrm{T}}, \ldots, \dot{q}_N^{\mathrm{T}}]^{\mathrm{T}}$. By using the properties of Lagrangian dynamics (Sect. 12.1) and noting that $P_{\mathcal{X}_i}(q_i)$ is a (globally) Lipschitz continuous function of q_i, we know that (12.2) is an autonomous system satisfying a Lipschitz condition. Then, consider the following Lyapunov function:

$$V(x) = \frac{1}{2}\sum_{i=1}^{N} \dot{q}_i^{\mathrm{T}} M_i(q_i)\dot{q}_i + \frac{1}{2}\sum_{i=1}^{N} \alpha_i \|q_i - P_{\mathcal{X}_i}(q_i)\|^2 + \frac{\beta}{4}\sum_{i=1}^{N}\sum_{j \in \mathcal{N}_i} a_{ij}\|q_i - q_j\|^2. \tag{12.3}$$

Based on the properties of Lagrangian dynamics (Sect. 12.1) and Assumption 12.1, it follows that $V(x)$ is radially unbounded, i.e., $V(x) \to \infty$ as $\|x\| \to \infty$. Therefore, the set $\Omega_c = \{x \in \mathbb{R}^{2Nn} | V(x) \le c\}$ is bounded, for all $c = V(x(0))$. The derivative of V along (3.5) with (12.1) is

$$\dot{V} = \sum_{i=1}^{N} \dot{q}_i^{\mathrm{T}}\left(\frac{1}{2}\dot{M}_i(q_i)\dot{q}_i + M_i(q_i)\ddot{q}_i\right) + \sum_{i=1}^{N} \alpha_i \dot{q}_i^{\mathrm{T}}(q_i - P_{\mathcal{X}_i}(q_i))$$

$$+ \frac{\beta}{2}\sum_{i=1}^{N}\sum_{j \in \mathcal{N}_i} a_{ij}(q_i - q_j)^{\mathrm{T}}(\dot{q}_i - \dot{q}_j)$$

$$= \sum_{i=1}^{N} \dot{q}_i^{\mathrm{T}}\left(-k_i\dot{q}_i - \alpha_i(q_i - P_{\mathcal{X}_i}(q_i)) - \beta \sum_{j \in \mathcal{N}_i} a_{ij}(q_i - q_j)\right)$$

$$+ \sum_{i=1}^{N} \alpha_i \dot{q}_i^{\mathrm{T}}(q_i - P_{\mathcal{X}_i}(q_i)) + \beta \sum_{i=1}^{N} \dot{q}_i^{\mathrm{T}} \sum_{j \in \mathcal{N}_i} a_{ij}(q_i - q_j)$$

$$= -\sum_{i=1}^{N} k_i \dot{q}_i^{\mathrm{T}}\dot{q}_i \le 0,$$

where we have used convex analysis to derive the first equality, the fact that $a_{ij} = a_{ji}$, and the second property of Lagrangian dynamics to derive the second equality. Note that $V(x)$ is continuously differentiable for all $x \in \mathbb{R}^{2Nn}$. Then, we take $\Omega = \Omega_c$ as the positively invariant compact set for Lemma 2.5. Then, based on LaSalle's Invariance Principle, we know that every solution of (12.2) converges to the set $\mathcal{M}$, where $\mathcal{M} = \{q_i \in \mathbb{R}^n, \dot{q}_i \in \mathbb{R}^n, \forall i \in \mathcal{V} \mid \dot{q} = 0, \text{ and } q, \dot{q} \text{ are subject to (12.2)}\}$.

Let $x(t)$ be a solution that belongs to $\mathcal{M}$. Then, based on (12.2), we know that

$$\dot{q} \equiv 0 \Rightarrow \alpha_i(q_i - P_{\mathcal{X}_i}(q_i)) + \beta \sum_{j \in \mathcal{N}_i} a_{ij}(q_i - q_j) \equiv 0, \ \forall i \in \mathcal{V}.$$

Pick any $q_0 \in \mathcal{X}_0$. Such a q_0 exists due to Assumption 12.1. Thus, it follows that for all $i \in \mathcal{V}$, $\beta(q_i - q_0)^\mathsf{T} \sum_{j \in N_i} a_{ij}(q_i - q_j) + \alpha_i(q_i - q_0)^\mathsf{T}(q_i - P_{\mathcal{X}_i}(q_i)) \equiv 0$. We then know that $\beta \sum_{i=1}^{N}(q_i - q_0)^\mathsf{T} \sum_{j \in \mathcal{N}_i} a_{ij}(q_i - q_j) + \alpha_i \sum_{i=1}^{N}(q_i - q_0)^\mathsf{T}(q_i - P_{\mathcal{X}_i}(q_i)) \equiv 0$.

It also follows that $\sum_{i=1}^{N}(q_i - q_0)^\mathsf{T} \sum_{j \in \mathcal{N}_i} a_{ij}(q_i - q_j)$
$= \frac{1}{2} \sum_{i=1}^{N} \sum_{j \in \mathcal{N}_i} a_{ij} \|q_i - q_j\|^2 \geq 0$ by noting that $a_{ij} = a_{ji}$. Also, we know that for all $i \in \mathcal{V}$, $(P_{\mathcal{X}_i}(q_i) - q_0)^\mathsf{T}(q_i - P_{\mathcal{X}_i}(q_i)) \geq 0$. It then follows that

$$(q_i - q_0)^\mathsf{T}(q_i - P_{\mathcal{X}_i}(q_i)) = \|q_i - P_{\mathcal{X}_i}(q_i)\|^2 + (P_{\mathcal{X}_i}(q_i) - q_0)^\mathsf{T}(q_i - P_{\mathcal{X}_i}(q_i))$$
$$\geq \|q_i - P_{\mathcal{X}_i}(q_i)\|^2.$$

This shows that $\sum_{i=1}^{N}(q_i - q_0)^\mathsf{T} \sum_{j \in \mathcal{N}_i} a_{ij}(q_i - q_j) \equiv 0$, and $\|q_i - P_{\mathcal{X}_i}(q_i)\| \equiv 0$, $\forall i \in \mathcal{V}$. Note that the above analysis holds for all $x(0) \in \mathbb{R}^{2Nn}$ and $\mathcal{D} = \mathbb{R}^{2Nn}$. Therefore, we know from LaSalle's Invariance Principle and the fact that $\mathcal{G}$ is connected that for all $q_i(0) \in \mathbb{R}^n$, $\dot{q}_i(0) \in \mathbb{R}^n$, $i \in \mathcal{V}$, $\lim_{t \to \infty} \dot{q}_i(t) = 0$, $\lim_{t \to \infty}(q_i(t) - P_{\mathcal{X}_i}(q_i(t))) = 0, \forall i \in \mathcal{V}$, and $\lim_{t \to \infty}(q_i(t) - q_j(t)) = 0, \forall i, j \in \mathcal{V}$. It then follows that for all $i \in \mathcal{V}$ and $l \in \mathcal{V}$

$$\|q_i - P_{\mathcal{X}_i}(q_i)\| \leq \|q_i - q_l\| + \|q_l - P_{\mathcal{X}_i}(q_l)\| + \|P_{\mathcal{X}_i}(q_l) - P_{\mathcal{X}_i}(q_i)\|$$
$$\leq 2\|q_i - q_l\| + \|q_l - P_{\mathcal{X}_i}(q_l)\|. \tag{12.4}$$

This implies that $\lim_{t \to \infty}(q_i(t) - P_{\mathcal{X}_i}(q_i(t))) = 0, \forall i \in \mathcal{V}$ and $l \in \mathcal{V}$. Therefore, $\lim_{t \to \infty} d(q_i(t), \mathcal{X}_0) = 0, \forall i \in \mathcal{V}$. This shows that cooperative set aggregation is achieved in the sense of Definition 12.1.

(Necessity.) Immediate from the fact that $\mathcal{X}_0$ is nonempty is a necessary condition such that the first part of Definition 12.1 can be achieved.

12.3 Switching Graphs

One issue of the communication unit is the possible communication link failure. The link failure becomes even more important when we consider the physical applications including controlling multiple autonomous vehicles in the environments with limited power. Therefore, it is necessary to consider the case of switching communication graphs. We associate the switching communication topology with a time-varying graph $\mathcal{G}_{\sigma(t)} = (\mathcal{V}, \mathcal{E}_{\sigma(t)})$.

The existence of the switching graph significantly complicates the problem. In order to simplify the problem, we assume that the exact information of system dynamical parameters is available and propose the following control:

$$\tau_i = C_i(q_i, \dot{q}_i)\dot{q}_i - kM_i(q_i)\dot{q}_i - \alpha_i M_i(q_i)(q_i - P_{\mathcal{X}_i}(q_i))$$

$$- \beta M_i(q_i) \sum_{j \in \mathcal{N}_i(\sigma(t))} a_{ij}(\sigma(t))(q_i - q_j), \quad \forall i \in \mathcal{V}, \qquad (12.5)$$

where $k > 0$ denotes generalized coordinate derivative damping, $\alpha_i > 0$ denotes the gain for the target set projection control, $\beta > 0$ denotes the gain for the cooperative control, and $a_{ij}(p)$ is the (i, j)th entry of the adjacency matrix A_p associated with graph $\mathcal{G}_p$, for all $p \in \mathcal{P}$.

Before moving on, the following fact has been shown in Proposition 4.10 of [16] and we restate it in the following lemma.

Lemma 12.3 *Consider a multi-agent system with the switching graph $\mathcal{G}_{\sigma(t)}$. The dynamics of node i is given by*

$$\dot{x}_i = - \sum_{j \in \mathcal{N}_i(\sigma(t))} a_{ij}(\sigma(t))(x_i - x_j) + \epsilon_i(t),$$

for all $i \in \mathcal{V}$, where ϵ_i is a piecewise continuous function. Suppose $\mathcal{G}_{\sigma(t)}$ is uniformly jointly quasi-strongly connected (see Definition 2.9) and $\lim_{t \to \infty} \epsilon_i(t) = 0$ for all $i \in \mathcal{V}$. Then,

$$\lim_{t \to \infty} (x_i(t) - x_j(t)) = 0, \quad \forall i, j \in \mathcal{V}.$$

The main result of this section is as follows.

Theorem 12.4 *Suppose that Assumption 12.1 holds, $\mathcal{X}_0$ is nonempty, and choose k sufficiently large. The multi-robotic system (3.5) with (12.5) achieves cooperative set aggregation in the sense of Definition 12.1 if $\mathcal{G}_{\sigma(t)}$ is uniformly jointly connected.*

Proof Note that closed-loop system of (3.5) and (12.5) can be written as

$$\ddot{q}_i = -k\dot{q}_i - \beta \sum_{j \in \mathcal{N}_i(\sigma(t))} a_{ij}(\sigma(t))(q_i - q_j) - \alpha_i(q_i - P_{\mathcal{X}_i}(q_i)). \qquad (12.6)$$

We next focus on the closed-loop system (12.6) and first show that $\lim_{t \to \infty} \left(q_i(t) - P_{\mathcal{X}_i}(q_i(t))\right) = 0$, and $\lim_{t \to \infty} \dot{q}_i(t) = 0$, for all $i \in \mathcal{V}$.

By picking any $q_0 \in \mathcal{X}_0$, we propose the following Lyapunov function:

$$V = \frac{1}{2} \sum_{i=1}^{N} \frac{1}{\beta} \dot{q}_i^{\mathrm{T}} \dot{q}_i + \sum_{i=1}^{N} \frac{1}{\beta}(q_i - q_0)^{\mathrm{T}} \dot{q}_i + \sum_{i=1}^{N} \frac{k}{2\beta} \|q_i - q_0\|^2$$

$$+ \frac{1}{2} \sum_{i=1}^{N} \frac{\alpha_i}{\beta} \| q_i - P_{\mathcal{X}_i}(q_i) \|^2,$$

where we choose $k > 1$ to guarantee V is positive definite. The derivative of V along (12.6) is

$$\dot{V} = \sum_{i=1}^{N} \frac{1}{\beta} \dot{q}_i^{\mathrm{T}} \left(-k\dot{q}_i - \beta \sum_{j \in \mathcal{N}_i(\sigma(t))} a_{ij}(\sigma(t))(q_i - q_j) - \alpha_i(q_i - P_{\mathcal{X}_i}(q_i)) \right)$$

$$+ \sum_{i=1}^{N} \frac{\alpha_i}{\beta} \dot{q}_i^{\mathrm{T}}(q_i - P_{\mathcal{X}_i}(q_i)) + \sum_{i=1}^{N} \frac{(q_i - q_0)^{\mathrm{T}}}{\beta} \left(-k\dot{q}_i - \beta \sum_{j \in \mathcal{N}_i(\sigma(t))} a_{ij}(\sigma(t)) \right)$$

$$\times (q_i - q_j) - \alpha_i(q_i - P_{\mathcal{X}_i}(q_i)) \Big) + \sum_{i=1}^{N} \frac{\|\dot{q}_i\|^2}{\beta} + \sum_{i=1}^{N} \frac{k}{\beta}(q_i - q_0)^{\mathrm{T}} \dot{q}_i$$

$$= -\sum_{i=1}^{N} \frac{k-1}{\beta} \|\dot{q}_i\|^2 - \sum_{i=1}^{N} \dot{q}_i^{\mathrm{T}} \sum_{j \in \mathcal{N}_i(\sigma(t))} a_{ij}(\sigma(t))(q_i - q_j) - \sum_{i=1}^{N} \frac{\alpha_i}{\beta}$$

$$\times (q_i - q_0)^{\mathrm{T}}(q_i - P_{\mathcal{X}_i}(q_i)) - \sum_{i=1}^{N}(q_i - q_0)^{\mathrm{T}} \sum_{j \in \mathcal{N}_i(\sigma(t))} a_{ij}(\sigma(t))(q_i - q_j).$$

It then follows that

$$\dot{V} \leq -\left[q(t) \ \dot{q}(t) \right] \left(\begin{bmatrix} L_{\sigma(t)} & \frac{L_{\sigma(t)}}{2} \\ \frac{L_{\sigma(t)}}{2} & K \end{bmatrix} \otimes I_n \right) \begin{bmatrix} q(t) \\ \dot{q}(t) \end{bmatrix}$$

$$- \sum_{i=1}^{N} \frac{\alpha_i}{\beta} d^2(q_i(t), \mathcal{X}_i) - \sum_{i=1}^{N} \frac{1}{\beta} \|\dot{q}_i(t)\|^2,$$

where $K = \frac{k-2}{\beta} I_N$, $q = [q_1^{\mathrm{T}}, q_2^{\mathrm{T}}, \ldots, q_N^{\mathrm{T}}]^{\mathrm{T}}$, $\dot{q} = [\dot{q}_1^{\mathrm{T}}, \dot{q}_2^{\mathrm{T}}, \ldots, \dot{q}_N^{\mathrm{T}}]^{\mathrm{T}}$, $L_{\sigma(t)}$ is the Laplacian matrix associated with graph $\mathcal{G}_{\sigma(t)}$ at time t, and we have used the fact that $(q_i - q_0)^{\mathrm{T}}(q_i - P_{\mathcal{X}_i}(q_i)) \geq \|q_i - P_{\mathcal{X}_i}(q_i)\|^2$ based on convex analysis.

It is trivial to show that L_p is symmetric and positive semi-definite, for all $p \in \mathcal{P}$. It follows that L_p can be diagonalized as $L_p = \Gamma_p^{-1} \Lambda_p \Gamma_p$, where Γ_p is a real orthogonal matrix, $\Lambda_p = \mathrm{diag}\{\lambda_p^1, \lambda_p^2, \ldots, \lambda_p^N\}$ and $\lambda_p^i \geq 0$ for all $i \in \mathcal{V}$ and all $p \in \mathcal{P}$. We then know that $F_p = \begin{bmatrix} \Gamma_p^{-1} & 0 \\ 0 & \Gamma_p^{-1} \end{bmatrix} P_p \begin{bmatrix} \Gamma_p & 0 \\ 0 & \Gamma_p \end{bmatrix}$, where $F_p = \begin{bmatrix} L_p & \frac{L_p}{2} \\ \frac{L_p}{2} & K \end{bmatrix}$ and $P_p = \begin{bmatrix} \Lambda_p & \frac{\Lambda_p}{2} \\ \frac{\Lambda_p}{2} & K \end{bmatrix}$. It then follows that the eigenvalue μ_p of P_p are the solutions of $\mu_p^2 - (\lambda_p^i + \frac{k-2}{\beta})\mu_p + \frac{k-2}{\beta}\lambda_p^i - \frac{1}{4}(\lambda_p^i)^2 = 0$ for all $p \in \mathcal{P}$. Thus, F_p is positive semi-definite for all $p \in \mathcal{P}$ if k is chosen such that $k \geq 2 + \frac{1}{4}\beta \max_{p \in \mathcal{P}}\{\lambda_{\max}(L_p)\}$. Since $\lambda_{\max}(L_p)$ can be bounded by $\lambda_{\max}(L_p) \leq 2\max_{i \in \mathcal{V}}\sum_{j \in \mathcal{N}_i(p)} a_{ij}(p)$ based on inequality (12) of [14], we know that it is sufficient to choose $k \geq 2 + \frac{(N-1)a^*\beta}{2}$ such that F_p is positive semi-definite for all $p \in \mathcal{P}$, where a^* is defined in Sect. 4.2. Under this condition, we know that

$$\dot{V} \leq -\sum_{i=1}^{N} \frac{\alpha_i}{\beta} d^2(q_i, \mathcal{X}_i) - \sum_{i=1}^{N} \frac{1}{\beta}\|\dot{q}_i\|^2 \leq 0. \tag{12.7}$$

Therefore, q_i and $\dot{q}_i$, $\forall i \in \mathcal{V}$, are bounded. We also know that (12.7) implies that $\int_0^\infty \left(\sum_{i=1}^{N} \frac{\alpha_i}{\beta} d^2(q_i(t), \mathcal{X}_i) + \sum_{i=1}^{N} \frac{1}{\beta}\|\dot{q}_i(t)\|^2\right) dt \leq V(0)$ is bounded. In addition, it follows that

$$\frac{d}{dt}\left(\sum_{i=1}^{N} \frac{\alpha_i}{\beta} d^2(q_i(t), \mathcal{X}_i) + \sum_{i=1}^{N} \frac{1}{\beta}\|\dot{q}_i(t)\|^2\right) = 2\sum_{i=1}^{N}\left(\frac{\alpha_i}{\beta}(q_i - P_{\mathcal{X}_i}(q_i))^{\mathsf{T}}\dot{q}_i + \frac{1}{\beta}\dot{q}_i^{\mathsf{T}}\ddot{q}_i\right).$$

Therefore, from (12.6) and the facts that q_i and $\dot{q}_i$, $\forall i \in \mathcal{V}$, are bounded, we know that

$$\frac{d}{dt}\left(\sum_{i=1}^{N} \frac{\alpha_i}{\beta} d^2(q_i(t), \mathcal{X}_i) + \sum_{i=1}^{N} \frac{1}{\beta}\|\dot{q}_i(t)\|^2\right)$$

is bounded $\forall t \geq 0$. Then, based on Barbalat's lemma [8], we can show that

$$\lim_{t \to \infty}\left(\sum_{i=1}^{N} \frac{\alpha_i}{\beta} d^2(q_i(t), \mathcal{X}_i) + \sum_{i=1}^{N} \frac{1}{\beta}\|\dot{q}_i(t)\|^2\right) = 0.$$

Therefore, $\lim_{t \to \infty} d(q_i(t), \mathcal{X}_i) = 0$, and $\lim_{t \to \infty} \dot{q}_i(t) = 0$, for all $i \in \mathcal{V}$. Finally, we know that $\lim_{t \to \infty}\left(q_i(t) - P_{\mathcal{X}_i}(q_i(t))\right) = 0$, for all $i \in \mathcal{V}$.

We next show $\lim_{t \to \infty}(q_i(t) - q_j(t)) = 0$, $\quad \forall i, j \in \mathcal{V}$. Define $x_i = q_i$, $x_{N+i} = q_i + \frac{1}{\beta}\dot{q}_i$, for all $i \in \mathcal{V}$. After some manipulations, (12.6) can be rewritten as

$$\dot{x}_i = -\beta(x_i - x_{N+i}), \tag{12.8a}$$

$$\dot{x}_{N+i} = -\sum_{j \in \mathcal{N}_i(\sigma(t))} a_{ij}(\sigma(t))(x_{N+i} - x_{N+j}) + \epsilon_i(t), \tag{12.8b}$$

where $i \in \mathcal{V}$, and $\epsilon_i = (1 - \frac{k}{\beta})\dot{q}_i + \frac{1}{\beta}\sum_{j \in \mathcal{N}_i(\sigma(t))} a_{ij}(\sigma(t))(\dot{q}_i - \dot{q}_j) - \frac{\alpha_i}{\beta}(q_i - P_{\mathcal{X}_i}(q_i))$, for all $i \in \mathcal{V}$. We have shown that $\lim_{t \to \infty} \epsilon_i(t) = 0$, for all $i \in \mathcal{V}$.

Consider system (12.8) as a new networked system with node set $\overrightarrow{\mathcal{V}} = \{1, 2, \ldots, 2N\}$. We associate this system with a new graph $\overrightarrow{\mathcal{G}}_{\sigma(t)} = (\overrightarrow{\mathcal{V}}, \overrightarrow{\mathcal{E}}_{\sigma(t)})$ and the corresponding neighbor set $\overrightarrow{\mathcal{N}}_i(\sigma(t))$ and adjacency matrix $\overrightarrow{A}_{\sigma(t)}$, where the connections and weights for agents $\{N + 1, N + 2, \ldots, 2N\}$ are defined by $\mathcal{E}_{\sigma(t)}$ and $A_{\sigma(t)}$. In addition, there exists persistent edge $a_{i(i+N)}(t) = \beta > 0$, for all $i = 1, 2, \ldots, N$ and all $t \geq 0$. Note that $\overrightarrow{\mathcal{G}}_{\sigma(t)}$ is switching from the sets of directed graphs. It is not hard to verify that $\overrightarrow{\mathcal{G}}_{\sigma(t)}$ is uniformly jointly quasi-strongly connected provided that $\mathcal{G}_{\sigma(t)}$ is uniformly jointly connected, and with the same uniform constant T.

It then follows from Lemma 12.3 that $\lim_{t\to\infty} \dot{q}_i(t) = 0$, $\lim_{t\to\infty}(q_i(t) - P_{\mathcal{X}_i}(q_i(t))) = 0$, $\forall i \in \mathcal{V}$, and $\lim_{t\to\infty}(q_i(t) - q_j(t)) = 0$. $\forall i, j \in \mathcal{V}$. Following (12.4) in the proof of Theorem 12.2, it follows that $\lim_{t\to\infty} d(q_i(t), \mathcal{X}_0) = 0$, $\forall i \in \mathcal{V}$. This shows that cooperative set aggregation is achieved in the sense of Definition 12.1.

12.4 Simulations

We now use numerical simulations to validate the effectiveness of the theoretical results obtained in Sects. 12.2 and 12.3. We assume that there are eight agents ($N = 8$) in the group. The system dynamics are given by [17], $\begin{bmatrix} M_{11,i} & M_{12,i} \\ M_{21,i} & M_{22,i} \end{bmatrix} \begin{bmatrix} \ddot{q}_{ix} \\ \ddot{q}_{iy} \end{bmatrix} +$ $\begin{bmatrix} C_{11,i} & C_{12,i} \\ C_{21,i} & C_{22,i} \end{bmatrix} \begin{bmatrix} \dot{q}_{ix} \\ \dot{q}_{iy} \end{bmatrix} = \begin{bmatrix} \tau_{ix} \\ \tau_{iy} \end{bmatrix}$, $i = 1, 2, \ldots, 8$, where $M_{11,i} = \theta_{1i} + 2\theta_{2i}\cos q_{iy}$, $M_{12,i} = M_{21,i} = \theta_{3i} + \theta_{2i}\cos q_{iy}$, $M_{22,i} = \theta_{3i}$, $C_{11,i} = -\theta_{2i}\sin q_{iy}\dot{q}_{iy}$, $C_{12,i} = -\theta_{2i}\sin q_{iy}(\dot{q}_{ix} + \dot{q}_{iy})$, $C_{21,i} = \theta_{2i}\sin q_{iy}\dot{q}_{ix}$, $C_{22,i} = 0$. Choose $\theta_{1i} = 1.301$, $\theta_{2i} = 0.256$, $\theta_{3i} = 0.096$, $i = 1, 2, \ldots, 8$.

12.4.1 Fixed Graphs

We first assume that the available target sets of all the agents are disks. The radii of the disks are $rl_i = 3$, $i = 1, 2 \ldots, 8$. Denote the coordinates of the center points as $l_i = [l_{ix}, l_{iy}]^T \in \mathbb{R}^2$, $i = 1, 2 \ldots, 8$ and $l_1 = l_3 = l_7 = [1.5, 1.5]^T$, $l_2 = l_6 = l_8 = [0, -3]^T$, and $l_4 = l_5 = [1.5, -3]^T$. The local projection term $P_{\mathcal{X}_i}(q_i)$ in control input (12.1) is specified as for each $i = 1, 2 \ldots, 8$,

$$P_{\mathcal{X}_i}(q_i) = \begin{cases} q_i, & \text{if } \|q_i - l_i\| \leq rl_i \\ q_{ir1}, & \text{if } \|q_i - l_i\| > rl_i \text{ and } q_{ix} > l_{ix} \\ q_{ir2}, & \text{otherwise,} \end{cases}$$

Fig. 12.3 The fixed graph $\mathcal{G}$

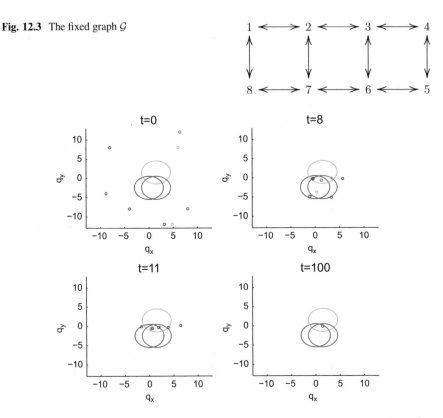

Fig. 12.4 Snapshots of the generalized coordinates of the multi-robotic system (3.5) with (12.1) for Sect. 12.2. The small circles denote the generalized coordinates of the agents and the large circles are target sets. As indicated by the plots, cooperative set aggregation is achieved

where $q_{ir1} = [l_{ix} + rl_i\cos\theta_i, l_{iy} + rl_i\sin\theta_i]$, $q_{ir2} = [l_{ix} + rl_i\cos(\pi + \theta_i), l_{iy} + rl_i\sin(\pi + \theta_i)]$, $\theta_i = \arctan\frac{q_{iy}-l_{iy}}{q_{ix}-l_{ix}}$. The initial states of the agents are given by $q_1(0) = [-8, 8]^T$, $q_2(0) = [6.4, 12]^T$, $q_3(0) = [-8, -8]^T$, $q_4(0) = [6, -8]^T$, $q_5(0) = [-8.8, -4]^T$, $q_6(0) = [4.8, -12]^T$, $q_7(0) = [-4, -8]^T$, $q_8(0) = [3.2, -12]^T$, $\dot{q}_1(0) = [-0.4, 0.4]^T$, $\dot{q}_2(0) = [0.8, -0.8]^T$, $\dot{q}_3(0) = [2.8, -2.8]^T$, and $\dot{q}_4(0) = [1.6, -1.6]^T$, $\dot{q}_5(0) = [-1.2, 0.8]^T$, $\dot{q}_6(0) = [1.6, -0.4]^T$, $\dot{q}_7(0) = [1.6, -2]^T$, and $\dot{q}_8(0) = [0.8, -0.8]^T$. The control parameters are chosen by $k_i = 1$, $\alpha_i = 1$, for all $i \in \mathcal{V}$, and $\beta = 1$. The graph $\mathcal{G}$ is given in Fig. 12.3. Also, the nonzero weight of adjacency matrix A associated with $\mathcal{G}$ is chosen to be 1, for all $i, j \in \mathcal{V}$.

For the multi-robotic system (3.5) with (12.1), snapshots of generalized coordinates and trajectories of both generalized coordinates and generalized coordinate derivatives of the agents are shown in Figs. 12.4 and 12.5. We see that all the generalized coordinates of agents converge to a common point in the intersection set of all the target sets $\mathcal{X}_i$, for all $i \in \mathcal{V}$, and the generalized coordinate derivatives of all the agents converge to zero. This agrees with the results of Theorem 12.2.

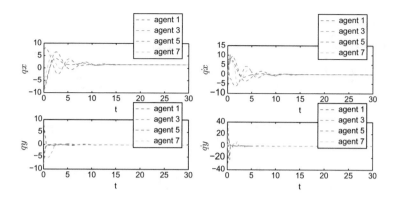

Fig. 12.5 The generalized coordinate and generalized coordinate derivative trajectories of the multi-robotic system (3.5) with (12.1) for Sect. 12.2

We next consider the case where $\mathcal{X}_i$ is non-convex set, for all $i \in \mathcal{V}$ to illustrate the importance of the convexity assumption in Assumption 12.1. We assume that the available target sets of all the agents have irregular forms as shown in Fig. 12.6. The initial states of the agents, the control parameters, the graph $\mathcal{G}$ are the same as those for the previous case.

For the multi-robotic system (3.5) with (12.1), snapshots of generalized coordinates and trajectories of generalized coordinate derivatives of the agents are shown in Figs. 12.6 and 12.7. We see that cooperative set aggregation cannot be achieved. All the agents neither converge into their target sets nor reach an agreement.

12.4.2 Switching Graphs

We finally consider the case of switching graphs. The nonzero weight of adjacency matrix $A_{\sigma(t)}$ of the generalized coordinates associated with $\mathcal{G}_{\sigma(t)}$ is chosen to be 1. The graph $\mathcal{G}_{\sigma(t)}$ switches between $\mathcal{G}_1$ (shown in Fig. 12.8) and $\mathcal{G}_2$ (shown in Fig. 12.9) at time instants $t_\ell = 5\ell$, $\ell = 0, 1, \ldots$. The control parameters are chosen as $k = 5$, $\beta = 1$, and $\alpha_i = 1$, for all $i \in \mathcal{V}$. The local sets are described by rectangles and presented in Fig. 12.10. The initial states of all agents are the same as those given in Sect. 12.4.1. The local projection term $P_{\mathcal{X}_i}(q_i)$ is specified as $P_{\mathcal{X}_i}(q_i) = [q_{irx}, q_{iry}]^T$ for each $i = 1, 2 \ldots, 8$, where

$$q_{irx} = \begin{cases} Lm_{ix}, & \text{if } q_{ix} < Lm_{ix} \\ LM_{ix}, & \text{if } q_{ix} > LM_{ix} \\ q_{ix}, & \text{otherwise,} \end{cases}$$

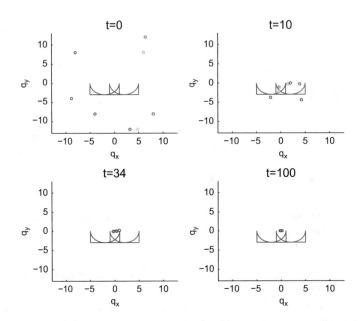

Fig. 12.6 Snapshots of the generalized coordinates of the multi-robotic system (3.5) with (12.1) for the case of non-convex target sets. As indicated by the plots, all the agents neither converge into their target sets nor reach an agreement

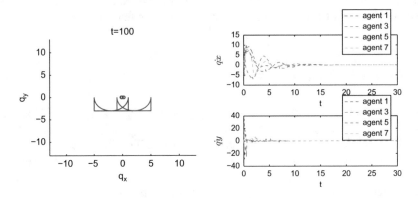

Fig. 12.7 Zoom-in of the final generalized coordinates and the generalized coordinate derivative trajectories of the multi-robotic system (3.5) with (12.1) for the case of non-convex target sets

Fig. 12.8 The graph $\mathcal{G}_1$

Fig. 12.9 The graph $\mathcal{G}_2$

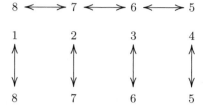

$$q_{iry} = \begin{cases} Lm_{iy}, & \text{if } q_{iy} < Lm_{iy} \\ LM_{iy}, & \text{if } q_{iy} > LM_{iy} \\ q_{iy}, & \text{otherwise.} \end{cases}$$

For the multi-robotic system (3.5) with (12.5), snapshots of generalized coordinates and trajectories of generalized coordinates and generalized coordinate derivatives of the agents are shown in Figs. 12.10 and 12.11. We see that cooperative set aggregation is achieved even when the graph is switching and the graph is disconnected at each time instant. This agrees with the results of Theorem 12.4.

12.5 Literature

The results in this chapter are based mainly on [12]. Cooperative control of multi-robotic networks is also studied in the existing literature. In particular, the author of [15] proposes distributed model-independent consensus algorithms for multiple Lagrangian systems in the leaderless setting. The coordination problem of multiple mechanical systems with safety guarantees is studied in [4]. The control laws are proposed to achieve both velocity synchronization and collision avoidance. The case of time-varying leader is studied in [5], where the nonlinear contraction analysis is introduced to obtain globally exponential convergence results. The connectivity maintenance problem is studied for multiple nonholonomic robotics in [6] and finite-time cooperative tracking algorithms are presented in [9] over graphs that are quasi-strongly connected. Distributed containment control is proposed in [10] and a sliding mode based strategy is introduced to estimate the leaders' generalized coordinate derivative information. A similar problem is also studied in [3], where continuous control algorithms are proposed to guarantee cooperative tracking with bounded errors. The authors of [11] consider a leader–follower coordinated tracking problem for multiple Lagrangian systems. A chattering-free algorithm with adaptive coupling gains is developed such that the tracking errors between the followers and

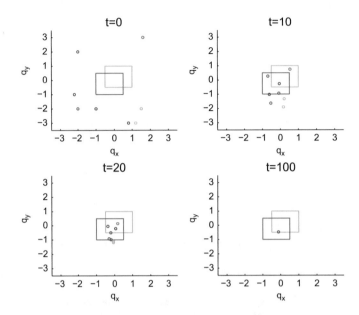

Fig. 12.10 Snapshots of the generalized coordinates of the multi-robotic system (3.5) with (12.5) for the case of switching communication graphs. The small circles denote the generalized coordinates of the agents and the large rectangles are target sets. As indicated by the plots, cooperative set aggregation is achieved

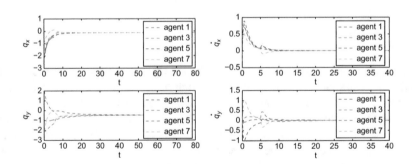

Fig. 12.11 Zoom-in of the final generalized coordinates and the trajectories of the generalized coordinates of the multi-robotic system (3.5) with (12.5) for the case of switching communication graphs

the leader are driven to zero. The influence of communication delays is discussed in [1, 13]. Sufficient conditions for reaching synchronization of multiple Lagrangian systems are established for the case of fixed and unknown delays in [13], and for the case of discontinuous time-varying delays in [1]. The region-based shape control is studied in [2, 7], where a group of robots modeled by Lagrangian dynamics is driven into a desired region while guaranteeing collision avoidance during the movement. A similar problem is studied in [18] and a multilevel structure approach is proposed such that the robots not only converge into the desired region, but also form a desired shape.

Acknowledgments ©2015 Elsevier. Reprinted, with permission, from Ziyang Meng, Tao Yang, Guodong Shi, Dimos V. Dimarogonas, Yiguang Hong, Karl H. Johansson, "Targeted agreement of cooperative Lagrangian systems", Automatica, vol. 84, pp. 109–116, 2017.

References

1. A. Abdessameud, I.G. Polushin, A. Tayebi, Synchronization of Lagrangian systems with irregular communication delays. IEEE Trans. Autom. Control **59**(1), 187–193 (2014)
2. C.C. Cheah, S.P. Hou, J.J.E. Slotine, Region-based shape control for a swarm of robots. Automatica **45**(10), 2406–2411 (2009)
3. G. Chen, F.L. Lewis, Distributed adaptive tracking control for synchronization of unknown networked Lagrangian systems. IEEE Trans. Syst. Man Cybern. B **41**(3), 805–816 (2011)
4. N. Chopra, D.M. Stipanovi, M.W. Spong, On synchronization and collision avoidance for mechanical systems, in *Proceedings of the American Control Conference*, pp. 3713–3718, Seattle, USA, 2008
5. S.J. Chung, U. Ahsun, J.J.E. Slotine, Cooperative robot control and concurrent synchronization of Lagrangian systems. IEEE Trans. Robot. **25**(3), 686–700 (2009)
6. D.V. Dimarogonas, K.J. Kyriakopoulos, Connectedness preserving distributed swarm aggregation for multiple kinematic robots. IEEE Trans. Robot. **24**(5), 1213–1223 (2008)
7. R. Haghighi, C.C. Cheah, Multi-group coordination control for robot swarms. Automatica **48**(10), 2526–2534 (2012)
8. H.K. Khalil, *Nonlinear Systems, Third Edition* (Prentice-Hall, USA, 2002)
9. S. Khoo, L. Xie, Z. Man, Robust finite-time consensus tracking algorithm for multirobot systems. IEEE/ASME Trans. Mechatron. **14**(2), 219–228 (2009)
10. J. Mei, W. Ren, G. Ma, Distributed containment control for Lagrangian networks with parametric uncertainties under a directed graph. Automatica **48**(4), 653–659 (2012)
11. Z. Meng, D.V. Dimarogonas, K.H. Johansson, Leader–follower coordinated tracking of multiple heterogeneous Lagrange systems using continuous control. IEEE Trans. Robot. **30**(3), 739–745 (2014)
12. Z. Meng, T. Yang, G. Shi, et al., Targeted agreement of multiple Lagrangian systems. Automatica **84**, 109–116 (2017)
13. E. Nuno, R. Ortega, L. Basanez, et al., Synchronization of networks of nonidentical Euler-Lagrange systems with uncertain parameters and communication delays. IEEE Trans. Autom. Control **56**(4), 935–941 (2011)
14. R. Olfati-Saber, J.A. Fax, R.M. Murray, Consensus and cooperation in networked multi-agent systems. Proc. IEEE **95**(1), 215–233 (2007)
15. W. Ren, Distributed leaderless consensus algorithms for networked Euler-Lagrange systems. Int. J. Control **82**(11), 2137–2149 (2009)

16. G. Shi, K.H. Johansson, Robust consensus for continuous-time multi-agent dynamics. SIAM J. Control Optim. **51**(5), 3673–3691 (2013)
17. M.W. Spong, S. Hutchinson, M. Vidyasagar, *Robot Dynamics and Control* (Wiley, 2006)
18. X. Yan, J. Chen, D. Sun, Multilevel-based topology design and shape control of robot swarms. Automatica **48**(12), 3122–3127 (2012)

Chapter 13
Energy Resource Coordination

The last application of networked dynamical systems comes from the power systems. Consider a power network of a large number of distributed energy resources. The optimal energy resource coordination problem is to minimize the total generation cost while meeting the total demand and satisfying individual generator output limits. Traditionally, this problem is solved by a centralized strategy. However, the centralized method requires a single control center that accesses the entire network's information and therefore may subject to performance limitations, such as high communication requirement and cost, substantial computational burden, and limited flexibility and scalability, and disrespect of privacy. It is thus desirable to develop distributed approaches to overcome these limitations and accommodate various resources in the future smart grid.

In particular, we focus on a power network consisting of distributed generators. We propose an algorithm based on the gradient push-sum method to solve the optimal energy resource coordination problem in a distributed manner over communication networks with switching communication topologies. This chapter shows that the proposed algorithm is guaranteed to solve the energy resource coordination problem if the switching communication topologies are uniformly jointly strongly connected. Numerical simulations are used to illustrate and validate the proposed algorithm.

13.1 Problem Formulation

In this chapter, we consider a power system consisting of distributed generators. An agent is assigned to each bus in the power system. The communication topology could be different from the physical network and is modelled as a switching graph $\mathcal{G}_k = (\mathcal{V}, \mathcal{E}_k)$, where the edge set changes over time due to unexpected loss of communication links. This section first presents the mathematical formulation of the energy resource coordination problem.

The goal of the optimal energy resource coordination problem is to minimize the total generation cost while meeting the total demand and satisfying individual generator output limits. The optimization objective is formulated in (13.1):

$$
\min_{x_i} \quad \sum_{i=1}^{N} C_i(x_i) \tag{13.1a}
$$

$$
\text{subject to} \quad \sum_{i=1}^{N} x_i = D, \tag{13.1b}
$$

$$
x_i \in \mathcal{X}_i := [x_i^{\min}, x_i^{\max}], \; i = 1, \ldots, N, \tag{13.1c}
$$

where N is the number of generators, x_i is the power generation of generator i, $C_i(\cdot) : \mathbb{R}_+ \to \mathbb{R}_+$ is the cost function of generator i, $x_i^{\min}$ and $x_i^{\max}$ are, respectively, the lower and upper bounds of the power generation of generator i, and D is the total demand satisfying $\sum_{i=1}^{N} x_i^{\min} \le D \le \sum_{i=1}^{N} x_i^{\max}$ in order to ensure the feasibility of problem (13.1).

We consider general convex cost functions that satisfy Assumption 13.1.

Assumption 13.1 *For each $i \in \{1, \ldots, N\}$, the cost function $C_i(\cdot) : \mathbb{R}_+ \to \mathbb{R}_+$ is strictly convex and continuously differentiable.*

13.2 Distributed Algorithm

This section proposes an algorithm that is capable to solve the optimal energy resource coordination problem in a distributed fashion over switching communication topologies.

First, we provide the centralized Lagrangian-based approach for solving the optimal energy resource coordination problem. Since (i) each cost function $C_i(\cdot)$ is convex, (ii) the constraint (13.1b) is affine, and (iii) the set $\mathcal{X}_1 \times \cdots \times \mathcal{X}_N$ is a polyhedral set, if we dualize problem (13.1) with respect to the constraint (13.1b), there is zero duality gap. Moreover, the dual optimal set is nonempty [2]. Consequently, solutions of the optimal energy resource coordination problem can be obtained by solving its dual problem.

For convenience, let $x = [x_1, \ldots, x_N]^\mathsf{T}$. Then, define the Lagrangian function

$$
\mathbf{L}(x, \lambda) = \sum_{i=1}^{N} C_i(x_i) - \lambda \left(\sum_{i=1}^{N} x_i - D \right).
$$

The corresponding Lagrange dual problem is

$$\max_{\lambda \in \mathbb{R}_+} \sum_{i=1}^{N} \Psi_i(\lambda) + \lambda D, \tag{13.2}$$

where

$$\Psi_i(\lambda) = \min_{x_i \in \mathcal{X}_i} C_i(x_i) - \lambda x_i. \tag{13.3}$$

Under Assumption 13.1, for any given $\lambda \in \mathbb{R}_+$, the right-hand side of (13.3) has a unique minimizer given by

$$x_i(\lambda) = \min\{\max\{\nabla C_i^{-1}(\lambda), x_i^{\min}\}, x_i^{\max}\}, \tag{13.4}$$

where ∇C_i^{-1} denotes the inverse function of ∇C_i, which exists over $[\nabla C_i(x_i^{\min}), \nabla C_i(x_i^{\max})]$, since ∇C_i is continuous and strictly increasing due to Assumption 13.1. Furthermore, there is at least one optimal solution to dual problem (13.2), and the unique optimal solution of the primal problem is given by

$$x_i^* = x_i(\lambda^*), \quad \forall i = 1, 2, \ldots, N, \tag{13.5}$$

where λ^* is any dual optimal solution.

For any given $\lambda \in \mathbb{R}_+$, because of the uniqueness of $x_i(\lambda)$, the dual function $\sum_{i=1}^{N} \Psi_i(\lambda) + \lambda D$ is differentiable at λ and its gradient is given by $-(\sum_{i=1}^{N} x_i(\lambda) - D)$ [1]. We can then apply the gradient method to solve the dual problem in (13.2):

$$\lambda(k+1) = \lambda(k) - \gamma(k) \left(\sum_{i=1}^{N} x_i(\lambda(k)) - D \right), \tag{13.6}$$

where $k \in \mathbb{Z}^+$ is the time step, $\lambda(0) \in \mathbb{R}$ can be arbitrarily assigned, and $\gamma(k)$ is the step-size at time step k. When designing a distributed algorithm based on (13.6), the main challenge is how to obtain the global quantity $\sum_{i=1}^{N} x_i(\lambda(k)) - D$ in a distributed manner. In this chapter, we will propose a distributed algorithm to avoid the need of the global quantity.

The dual problem in (13.2) can be converted into

$$\max_{\lambda \in \mathbb{R}} \sum_{i=1}^{N} \Phi_i(\lambda), \tag{13.7}$$

where

$$\Phi_i(\lambda) = \min_{x_i \in \mathcal{X}_i} C_i(x_i) - \lambda(x_i - D_i), \tag{13.8}$$

and D_i is a virtual local demand at each bus such that $\sum_{i=1}^{N} D_i = D$. The gradient of $\Phi_i(\lambda)$ is

$$\nabla \Phi_i(\lambda) = -(x_i(\lambda) - D_i). \tag{13.9}$$

Motivated by the gradient push-sum method [8], a distributed algorithm is proposed in algorithm (13.10) to solve the equivalent dual problem (13.7). In the proposed algorithm, each agent i maintains scalar variables $v_i(k)$, $w_i(k)$, $y_i(k)$, $\lambda_i(k)$, $x_i(k)$, where $x_i(k)$ and $\lambda_i(k)$ are estimations of the primal and dual optimal solution, respectively. At each time step k, each agent $i \in \mathcal{V}$ updates its variables according to (13.10).

$$w_i(k+1) = \frac{v_i(k)}{d_i(k)+1} + \sum_{j \in \mathcal{N}_i^{\text{in}}(k)} \frac{v_j(k)}{d_j(k)+1}, \tag{13.10a}$$

$$y_i(k+1) = \frac{y_i(k)}{d_i(k)+1} + \sum_{j \in \mathcal{N}_i^{\text{in}}(k)} \frac{y_j(k)}{d_j(k)+1}, \tag{13.10b}$$

$$\lambda_i(k+1) = \frac{w_i(k+1)}{y_i(k+1)}, \tag{13.10c}$$

$$x_i(k+1) = \min\{\max\{\nabla C_i^{-1}(\lambda_i(k+1)), x_i^{\min}\}, x_i^{\max}\}, \tag{13.10d}$$

$$v_i(k+1) = w_i(k+1) - \gamma(k+1)(x_i(k+1) - D_i). \tag{13.10e}$$

The step-size $\gamma(k+1)$ satisfies the following decay conditions:

$$\sum_{k=1}^{\infty} \gamma(k) = \infty, \quad \sum_{k=1}^{\infty} \gamma^2(k) < \infty,$$

$$\gamma(k) \le \gamma(s) \text{ for all } k > s \ge 1. \tag{13.11}$$

One typical selection is $\gamma(k) = \frac{a}{k+b}$, where $a > 0$ and $b \ge 0$. In this algorithm, each agent i needs to know its out-degree $d_i(k)$ and sends the quantities $\frac{v_i(k)}{d_i(k)+1}$ and $\frac{y_i(k)}{d_i(k)+1}$ to all the agents j in its out-neighbors set. In initialization step, $v_i(0)$ is assigned with an arbitrary value and $y_i(0) = 1$ for all $i \in \mathcal{V}$.

According to (13.9), $-(x_i(k+1) - D_i)$ in (13.10e) is the gradient of the function $\Phi_i(\lambda)$ at $\lambda = \lambda_i(k+1)$. Without (13.10d) and the gradient term in (13.10e), the algorithm would be reduced to a particular version of push-sum algorithm [7], or ratio consensus algorithm [3, 4] for computing the average of initial values in directed graphs. In this case, all $\lambda_i(k+1)$ converge to a common value. The inclusion of the gradient term in the update of $v_i(k+1)$ is to ensure that all $\lambda_i(k+1)$ converge to the optimal incremental cost λ^*.

We next show that the proposed distributed algorithm (13.10) is capable to solve the optimal energy resource coordination problem over switching graphs that satisfy Assumption 13.2, as stated in Theorem 13.1.

Assumption 13.2 *The switching graph $\mathcal{G}_k$ is uniformly jointly strongly connected (see Definition 3.1).*

Theorem 13.1 *Under Assumptions 13.1 and 13.2, distributed algorithm given in (13.10) with the step-size $\gamma(k)$ satisfying conditions in (13.11) solves the optimal energy resource coordination problem, i.e., $\lambda_i(k) \to \lambda^*$, and $x_i(k) \to x_i^*$ as $k \to \infty$ for all $i \in \mathcal{V}$.*

Proof Note that the equivalent dual problem (13.7) has the same form as the optimization problem considered in [8]. The only difference is that the dual problem is a maximization problem, while the problem in [8] is a minimization problem. In order to apply [8, Theorem 1] to show Theorem 13.1, we need to verify that all the conditions are satisfied:

- The condition (a) is that the graph is uniformly jointly strongly connected, which is satisfied in our case as assumed in Assumption 13.2.
- The condition (b) is that each function in the minimization problem is convex and the optimal set is nonempty. This is also satisfied in our case since each function $\Phi_i(\lambda)$ in the maximization problem (13.7) is concave and the optimal set is nonempty, which is guaranteed by Assumption 13.1.
- The condition (c) is that the (sub)gradient of each function in the problem is uniformly bounded. This is indeed satisfied in our case since it follows from (13.9) that the gradient of each function $\Phi_i(\lambda)$ is uniformly bounded, i.e.,

$$\left| \nabla \Phi_i(\lambda_i(k+1)) \right| = \left| -(x_i(\lambda_i(k+1)) - D_i) \right|$$

$$\leq \max_{i \in \mathcal{V}} x_i^{\max} + \max_{i \in \mathcal{V}} D_i. \tag{13.12}$$

Therefore, all the conditions are satisfied and the result follows.

13.3 Simulations

In this section, we use numerical simulations to illustrate and validate the proposed algorithm. We adopt the 4-bus system where the communication topology is switching [10] for comparison purpose. Four generators are selected from three types, and the power output ranges and parameters of quadratic cost functions for each generator type are given in Table 13.1. The communication topology is modelled as a switching graph $\mathcal{G}_k$ switching among three fixed topologies $\mathcal{G}_1, \mathcal{G}_2,$ and $\mathcal{G}_3$ shown in Fig. 13.1 at each time step. In particular,

Table 13.1 Generator parameters

Type	A (Gen. 1&2)	B (Gen. 3)	C (Gen. 4)
Range (MW)	[150,600]	[100,400]	[50,200]
a_i ($/MW^{2}h)	0.00142	0.00194	0.00482
b_i ($/MWh)	7.2	7.85	7.97
c_i ($/h)	510	310	78

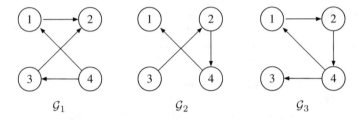

Fig. 13.1 Switching communication topology

$$
\mathcal{G}_k = \begin{cases}
\mathcal{G}_1, & \text{if } k \in [0, 1) \cup [3, 4) \cdots \cup [3s, 3s + 1) \ldots, \\
\mathcal{G}_2, & \text{if } k \in [1, 2) \cup [4, 5) \cdots \cup [3s + 1, 3s + 2) \ldots, \\
\mathcal{G}_3, & \text{if } k \in [2, 3) \cup [5, 6) \cdots \cup [3s + 2, 3s + 3) \ldots,
\end{cases}
$$

where $s \in \mathbb{Z}_+$. It is easy to check that each of the fixed topologies $\mathcal{G}_1, \mathcal{G}_2$, and $\mathcal{G}_3$ is not strongly connected. For example, there is no directed path from agent 2 to agent 4 in $\mathcal{G}_1$. However, the switching graph $\mathcal{G}_k$ is uniformly jointly strongly connected since the joint graph $\mathcal{G}([k, k + T))$ is strongly connected for any $k \geq 0$ with $T = 3$. Thus, Assumption 13.2 is satisfied with $T = 3$. According to Theorem 13.1, the proposed algorithm solves the optimal energy resource coordination problem over the switching graph. To implement the proposed algorithm, we first choose the virtual local demands at each bus as $D_1 = 500$ MW, $D_2 = 500$ MW, $D_3 = 350$ MW, and $D_4 = 150$ MW. The total demand is $D = \sum_{i=1}^{4} D_i = 1500$ MW, which is unknown to the agent at each bus.

With a step-size of $\gamma(k) = \frac{0.01}{k}$, the simulation results are shown in Fig. 13.2. As can be seen, $\lambda_i(k)$ converge to the optimal incremental cost $\lambda^* = 8.84$ $/MWh, and x_i converges to the optimal generation $x_1^* = 577.46$ MW, $x_2^* = 577.46$ MW, $x_3^* = 255.16$ MW, and $x_4^* = 90.25$ MW as shown in Fig. 13.2b, which agree with the centralized solution and the one obtained in [10]. In particular, at time step $k = 250$, the maximum difference between all the λ_i is 0.0126 $/MWh, which is quite small. The total generation gradually meets the total demand 1500 MW.

Fig. 13.2 Simulation results for the system of four generators over switching communication topologies. (**a**) Incremental cost ($/MWh). (**b**) Generation (MW) . (**c**) Total generation vs. demand (MW)

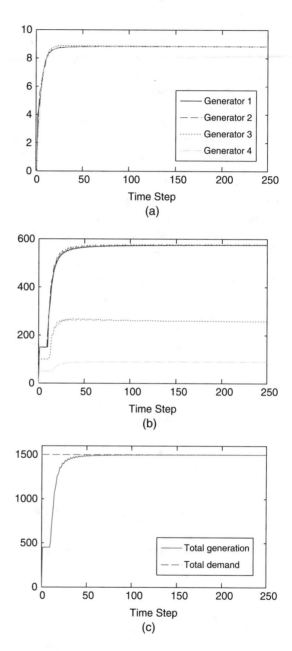

13.4 Literature

The results in this chapter are based mainly on [13]. For further results on distributed algorithms for energy resource coordination problem, see., e.g., [5, 6, 10–12, 14, 15].

In particular, the authors of [14] propose a leader–follower consensus-based algorithm where the leader collects the mismatch between demand and generation and then leads the updates of marginal cost in the system. To avoid the requirement of a leader, a two-level consensus-based algorithm is proposed in [15], where the upper level is the consensus and gradient algorithm, and the lower level executes the classical consensus by choosing the local mismatch as the consensus variable. In the algorithm proposed in [6], in addition to consensus part, an innovation term is introduced to ensure the balance between system generation and demand. All these three algorithms are only applicable to undirected communication topologies, i.e., the information must be exchanged bidirectionally. For the directed communication graphs, the authors of [5] propose a ratio consensus-based algorithm that relies on two linear iterations, and the authors of [11] estimate the mismatch with all the agents being participated. The authors of [11] propose minimum-time consensus-based algorithm to solve the optimal energy resource coordination problem in a minimum number of time steps. As for generation cost functions, most of the existing studies assume quadratic functions, whereas [10] considers general convex functions. For the switching communication topologies, the authors of [10] propose a non-negative-surplus based distributed algorithm to solve the EDP over switching graphs but without time delays. For the case where the communication links are subject to packet drops, the authors of [9] develop a distributed algorithm based on the push-sum method and the running sum method.

Acknowledgments ©2017 IEEE. Reprinted, with permission, from Tao Yang, Jie Lu, Di Wu, Junfeng Wu, Guodong Shi, Ziyang Meng, Karl H. Johansson, "Distributed algorithm for economic dispatch over time-varying directed networks with delays", IEEE Transactions on Industrial Electronics, vol. 64, no. 6, pp. 5095–5106, 2017.

References

1. D.P. Bertsekas, *Nonlinear Programming* (Athena Scientific, Belmont, USA, 1999)
2. D.P. Bertsekas, A. Nedić, A. Ozdaglar, *Convex Analysis and Optimization* (Athena Scientific, Belmont, MA, 2003)
3. T. Charalambous, Y. Yuan, T. Yang, W. Pan, C.N. Hadjicostis, M. Johansson, Distributed finite-time average consensus in digraphs in the presence of time-delays. IEEE Trans. Control Netw. Syst. **2**(4), 370–381 (2015)
4. A.D. Domínguez-García, C.N. Hadjicostis, Distributed strategies for average consensus in directed graphs, in *Proceedings of the 50th IEEE Conference Decision and Control*, pp. 2124–2129, Orlando, USA, 2011
5. A.D. Domínguez-García, S.T. Cady, C.N. Hadjicostis, Decentralized optimal dispatch of distributed energy resources, in *Proceedings of the IEEE Conference on Decision and Control* pp. 3688–3693, Maui, USA, 2012
6. S. Kar, G. Hug, Distributed robust economic dispatch in power systems: a consensus + innovations approach, in *Proceedings of the IEEE Power and Energy Society General Meeting*, pp. 1–8, San Diego, USA, 2012
7. D. Kempe, A. Dobra, J. Gehrke, Gossip-based computation of aggregate information, in *Proceedings of the Annual IEEE Symposium on Foundations of Computer Science*, pp. 482–491, Cambridge, USA, 2003

8. A. Nedić, A. Olshevsky, Distributed optimization over time-varying directed graphs. IEEE Trans. Autom. Control **60**(3), 601–615 (2015)

9. J. Wu, T. Yang, D. Wu, et al., Distributed optimal dispatch of distributed energy resources over lossy communication networks. IEEE Trans. Smart Grid **8**(6), 3125–3137 (2017)

10. Y. Xu, K. Cai, T. Han, et al., A fully distributed approach to resource allocation problem under directed and switching topologies, in *Proceedings of the Asian Control Conference*, pp. 1–6, Kota Kinabalu, Malaysia, 2015

11. S. Yang, S. Tan, J.X. Xu, Consensus based approach for economic dispatch problem in a smart grid. IEEE Trans. Power Syst. **28**(4), 4416–4426 (2013)

12. T. Yang, D. Wu, Y. Sun, et al., Minimum-time consensus based approach for power system applications. IEEE Trans. Ind. Electron. **63**(2), 1318–1328 (2016)

13. T. Yang, J. Lu, D. Wu, et al., A distributed algorithm for economic dispatch over time-varying directed networks with delays. IEEE Trans. Ind. Electron. **64**(6), 5095–5106 (2017)

14. Z. Zhang, M.Y. Chow, Convergence analysis of the incremental cost consensus algorithm under different communication network topologies in a smart grid. IEEE Trans. Power Syst. **27**(4), 1761–1768 (2012)

15. Z. Zhang, X. Ying, M.Y. Chow, Decentralizing the economic dispatch problem using a two-level incremental cost consensus algorithm in a smart grid environment, in *Proceedings of the North American Power Symposium*, pp. 1–7, Boston, USA, 2011

Printed in the United States
by Baker & Taylor Publisher Services